Routledge Introductions to Development
Series Editors:
John Bale and David Drakakis-Smith

Tropical Africa

Tropical Africa occupies a marginal position in the world economic system yet it covers an area larger than North and Central America. The region has seemingly insurmountable problems. An inadequate understanding of its own peoples and environments has led to inappropriate development strategies and projects often causing more problems than solutions.

Tropical Africa reveals the region's diversity and dynamism, showing how its peoples and environments have interacted through time and space. Considerable social, economic and political instability still exists since independence; this book considers development strategies, the environment and cooperation between and within African states in the light of past, present and future problems. Detailed case studies bring these patterns and processes to life.

Tony Binns is Lecturer in Geography at the School of African and Asian Studies, University of Sussex.

In the same series

John Cole
Development and Underdevelopment
A Profile of the Third World

David Drakakis-Smith
The Third World City

Allan and Anne Findlay
Population and Development in the Third World

Avijit Gupta
Ecology and Development in the Third World

John Lea
Tourism and Development in the Third World

John Soussan
Primary Resources and Energy in the Third World

Chris Dixon
Rural Development in the Third World

Alan Gilbert
Latin America

Janet Henshall Momsen
Women and Development in the Third World

David Drakakis-Smith
Pacific Asia

Rajesh Chandra
Industrialization and Development in the Third World

Tony Binns

Tropical Africa

Routledge
Taylor & Francis Group

LONDON AND NEW YORK

First published 1994
by Routledge
2 Park Square, Milton Park, Abingdon, Oxfordshire OX14 4RN

Simultaneously published in the USA and Canada
by Routledge
711 Third Avenue, New York, NY 10017

First issued in hardback 2015

Routledge is an imprint of the Taylor and Francis Group, an informa business

© 1994 Tony Binns

Typeset in Times by Florencetype Ltd, Kewstoke, Avon

British Library Cataloguing in Publication Data
A catalogue record for this book is available from the British Library

Library of Congress Cataloging in Publication Data
has been applied for

ISBN 13: 978-1-138-83497-2 (hbk)
ISBN 13: 978-0-415-04801-9 (pbk)

Contents

Plates

Figures

Tables

Introduction

Africa has fascinated me for almost a quarter of a century. My regular research visits and numerous geography field courses with sixth formers and undergraduates have further strengthened my interest. Longer periods spent teaching in African universities have enabled me to get to know many Africans well and get to grips with some of the more subtle aspects of politics and culture which are often not apparent on shorter visits. It is the sheer complexity of Africa that has fired my continuing interest.

Twenty-five years of studying Africa have left two dominant impressions uppermost in my mind. First, that the quality of life does not seem to have improved much during this time, at least in the rural areas with which I am more familiar. Rural villages still look as poor as ever, with a predominance of mud-built houses and an absence of basic amenities such as piped water, proper sanitation and electricity. Rural people barely live longer than they did twenty-five years ago, infant mortality is shamefully high, households have few material possessions and children with ragged clothing and bare feet are commonplace. What has gone wrong, one might ask, since those heady days of independence? My second lasting impression, and without intending to be either trivial or patronising, is that although so many people are materially poor in Africa, their warmth, friendliness and optimism far outweigh anything one might experience at home. The media frequently portray Africa as a depressing place, full of problems and populated with starving and hopeless people. Fortunately, such images are restricted to relatively small areas where events such as drought and civil war have destroyed livelihoods. The media feeds on such sorrowful images.

This book, which is based on my experiences of tramping and teaching about Africa, attempts to reveal the region's diversity and dynamism and show how its peoples and environments have interacted through time and space. In view of the vastness of the African continent, the wealth of its subject matter and the limited space available here, I have decided to focus on the forty-three mainland and island states which comprise tropical Africa. I have therefore excluded the states north of the Sahara and most of the countries of southern Africa. Tropical Africa covers an area of almost 22 million square kilometres, larger than the combined areas of North and Central America. I have tried to be positive in my approach, recognizing the impressive achievements of Africa's peoples, whilst not being so blinkered as to ignore the region's many seemingly insurmountable problems and the marginal position that Africa currently occupies in the world economic system. Wherever possible, detailed case study material has been used to bring life to patterns, processes and issues in an attempt to overcome the many problems associated with broad generalizations which are all too frequently made when detailed understanding is lacking.

Writing the book has on the whole been enjoyable, giving me an opportunity to reflect on my experiences and bring together some of my earlier work. I am entirely responsible for any errors of detail. Many people have helped to make the research and writing both stimulating and relatively pain-free. Susan Rowland at Sussex University has worked hard to produce the maps and diagrams from my rough drafts, whilst Jane Surry has assisted with the drawing-up of tables and endless photocopying. The book is dedicated to my family. My two children, Sarah and Joseph, have been particularly understanding and have coped with my frequent absences, knowing that once the work is finished we will visit southern Africa together! Last, but by no means least, my wife Margaret has sought out often obscure source material to complete the case studies and has given me much encouragement to finish the job.

Tony Binns
Hove, Sussex
January 1993

1
Tropical Africa: continuity and change

Images of Africa

The popular image of Africa is still that of 'the dark continent', as it was first portrayed by nineteenth-century explorers such as Stanley and Livingstone. Africa, and particularly tropical Africa, is often perceived as being little-known and mysterious, populated by hostile 'natives' and wild animals, where travel is hindered by harsh environments such as vast deserts and impenetrable jungles. More recently, tropical Africa has become synonymous with famine, drought, poverty and many other problems, a region characterized by little progress in economic, social or cultural development. The root cause of tropical Africa's problems, it is commonly suggested, is environmental factors, in conjunction, to a greater or lesser degree, with political and economic instability. Africa is seen as a predominantly rural continent, where an inability to feed its growing population is due to inefficient and outdated farming systems operated by an inadequately trained and poorly motivated workforce who are reluctant to adopt modern methods. Since independence, large-scale imports of food have been necessary, together with a multitude of aid programmes sponsored by governments and charity organizations in the developed world. In short, Africa is still perceived as being primitive, backward and poor, and furthermore, these characterisics, it is often suggested, are of its own making.

Such generalized images and stereotypes unfortunately ignore the great physical and human diversity of the African continent and fail to appreciate the complex historical processes which underlie this diversity. Africa is a vast continent, second only in size to Asia, stretching 8320 km from Tangiers in the north to Cape Town in the south, and 7360 km from Dakar in the west to Cape Guardafui, the easternmost point of the African horn. Once the cradle of the world's earliest civilizations, Africa now has some 500 million people, comprising a wide range of ethnic, language and religious groups. The continent's pre-colonial history was rich, varied and often highly sophisticated. It is only in the last hundred years or so that colonialism has transformed economies and societies and has pulled Africa, sometimes unwillingly, into the world economic system through trade in crops, minerals and other resources.

Understanding Africa

Longstanding myths and stereotypes about Africa, built up since the first Europeans set foot on the continent, and continually portrayed in the media, including school textbooks, are difficult to eradicate. These perceptions, often based upon an inadequate understanding of African environments, societies, cultures and economies, have sometimes, directly or indirectly, compounded Africa's problems. There are many examples of this, such as the shape of countries, the alignment of boundaries and the ethnic composition of African states. The 'great powers', meeting in Berlin in 1885 to divide up the African 'cake', showed little concern for the future governability and development of African countries and peoples. The Europeans merely staked their claims, with little consideration for the consequences. The result was the formation of countries such as the British colony of The Gambia (see Case study A). In some cases the colonial powers established boundaries, often straight lines, which showed little regard for topography or traditional tribal areas. The boundary between Ghana and Togo in West Africa, for example, divides the homeland of the Ewe people. Algeria, Libya, Namibia and Egypt each have at least one straight-line boundary, whilst in the case of Tanzania and Kenya, the straight-line border bends around Mount Kilimanjaro to include the mountain in Tanzania.

In the carving-up of Africa amongst the Europeans in the late-nineteenth century, states were created virtually overnight and disparate peoples thrown together. This led to frequent problems in both colonial and post-colonial periods as in the case of Nigeria, Africa's most populous state. Nigeria has almost 400 language groups and many

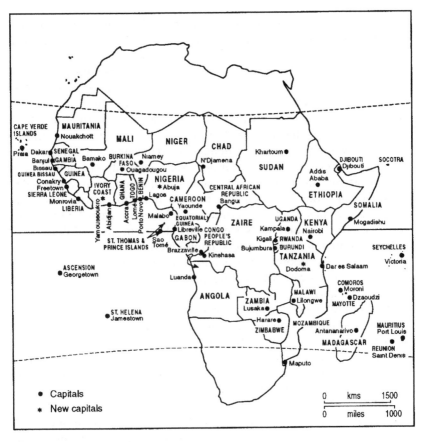

Figure 1.1 Countries and capitals 1992

tribal groups, of which the Hausa, Fulani, Yoruba and Ibo are the largest. Northern Nigeria is dominated by the Hausa and Fulani and is predominantly Muslim, whilst the south, dominated by the Yoruba and Ibo, is mainly Christian. Since independence from Britain in 1960, Nigeria has had two civilian and six military governments, five *coups d'état* and a bloody civil war (1967–70). Independent Nigeria has experienced great difficulty in holding important population censuses and surveys, a much-needed national census being at last undertaken in November 1991. In the early 1990s the country has been searching deeply to find the most appropriate form of future government, and its military leader, President Babangida, announced a return to civilian democratic rule in 1993. Nigeria's recurring political, religious and tribal rivalries have led some people to ask whether in fact the country will ever be governable.

Misunderstandings of a rather different nature are apparent in the ways that traditional farming and pastoral systems have been viewed and treated in colonial times and since independence. Traditional African farming systems, such as shifting cultivation and rotational bush fallow, have often been accused of being of low productivity, utilizing primitive methods, being environmentally damaging and unresponsive to market demands. African farmers are commonly characterized as being irrational and unaware of economic trends and more concerned to satisfy family food requirements than to maximize agricultural production. Such observations are frequently based on partial understanding, perhaps examining the systems only at certain times of the year or neglecting to talk with farmers themselves. Often in such surveys, remoter villages, poorer families and women are ignored because it is easier to communicate with literate male farmers living near a main road and close to a town.

Agricultural policies in African countries have favoured a 'technological transformation' approach, with large-scale development schemes, heavy use of machinery, fertilizers, pesticides and high-yielding crop varieties. The continent is littered with broken tractors and inappropriate development schemes. Recent research suggests that a better approach to increasing the productivity of African farming systems would be to harness farmers' considerable knowledge by introducing small-scale projects closely geared to the needs and aims of individuals or groups. Greater use of local knowledge, raw materials and intermediate technology is thought to be generally more appropriate than a 'high-tech' approach.

Similar misunderstandings surround pastoral systems in marginal arid and semi-arid areas. Governments are often critical of regular migrations of people and livestock in search of pasture and water. Such migrations may result in the spread of disease, tax evasion and clashes with settled cultivators. Practices such as burning, to stimulate new grass growth, and overgrazing are seen as environmentally destructive and linked with processes such as desertification. Governments frequently adopt a negative attitude towards pastoralists, believing that the quality of livestock and pasture can only be improved with settlement and the development of commercial ranching. Such schemes, however, have had little success in Africa, suggesting that there is a need to understand traditional pastoral systems more fully and perhaps to try to improve them with the help of herdsmen. It should be appreciated that mobility is an important way of coping with the ever-present threat of drought. The restriction of pastoralists' mobility and the loss of land to the cultivation of cash crops, such as groundnuts and cotton, has pushed herders into even more marginal areas where vulnerability to drought is much greater. African pastoralists have a wealth of environmental knowledge and their lifestyles may actually be the most effective and efficient given the harshness of climate and the poverty of environmental resources.

Criticism may also be levelled at certain government officials and so-called development 'experts' in the colonial and post-colonial periods for their lack of detailed understanding of the people with whom they were dealing. If Africa's economies and people are to develop in the future, much more emphasis should be placed on traditional ways of life and the wealth of knowledge and understanding which ordinary Africans possess. The dictatorial 'transformation approach' of the past must be replaced with a more democratic form of development which has indigenous Africans, whether rural or urban, as its main focus and genuinely seeks to fulfil their needs and ambitions.

Africa's colonial legacy

The colonial period was a relatively brief interlude in African history, yet its impact was profound and debate continues over the merits and problems of colonialism. The colonial powers demarcated boundaries and introduced administrative, legal, health, education and transport systems modelled on their own.

Economically, colonialism programmed African countries to consume what they do not produce and to produce what they do not consume. Africa was brought into the world economic system with the large-scale production and export of agricultural and mineral resources. The marketing of commodities by trading companies and multi-national organizations on a global scale developed under colonialism. The production of cash crops for export took priority over domestic food crops, whilst in the mining and industrial sectors, emphasis was placed on the extraction of minerals and their export in an unprocessed state. Manufacturing industries were discouraged, since their products might compete with European manufactured goods. Within African countries, traditional power structures were subordinated to colonial structures,

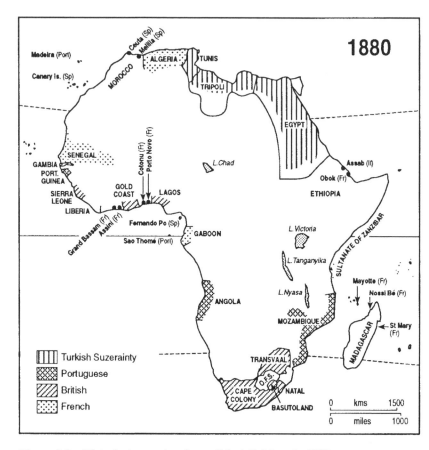

Figure 1.2a Historical map showing political divisions in 1880

with expatriate governors and district officers making key decisions that affected people's everyday lives. In many territories, the Colonial Governor ruled by decree or proclamation, was deferred to by all and lived in sumptuous surroundings. The perceptions and aspirations of individual Africans were transformed as a result of their contact with foreign people, foreign systems, foreign structures and foreign ways of life.

Even before the massive distribution of land resulting from the Berlin Conference of 1884–5, some of the continent had already passed under the control of the imperial powers (see Figure 1.2a). Treaties were signed

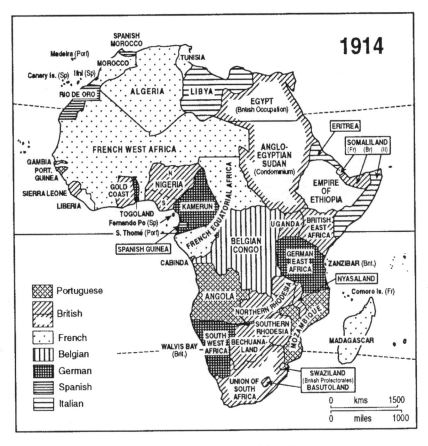

Figure 1.2b Historical map showing political divisions in 1914

with various African leaders, sometimes the latter not understanding the full implications of what they had signed. In other cases, where African peoples were less willing to cooperate, expensive and often bloody military campaigns were conducted against them. Occupation and pacification were followed by control and administration. Traders and missionaries often penetrated the countries first, whilst the tentacles of formal administration were gradually extended into the remotest areas.

By 1914 the political map of colonial Africa was complete and the colonial powers were unfolding their various policies and programmes. Approaches to colonial rule differed amongst the European powers.

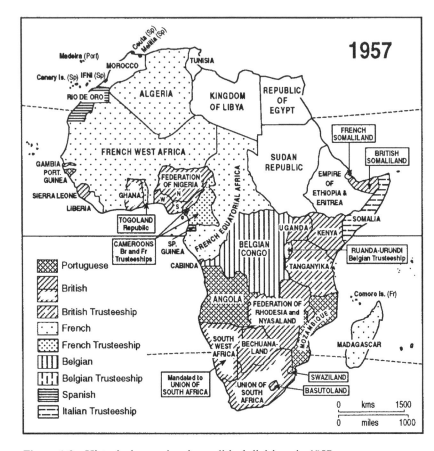

Figure 1.2c Historical map showing political divisions in 1957

The British adopted a pragmatic and decentralized approach, the essence of which was 'indirect rule', whereby Britain maintained indigenous cultures and societies by working wherever possible through local social and political systems, such as the Emirs in northern Nigeria. The British respected local institutions and feared the problems which might follow their decay. Under indirect rule, large areas could be administered by relatively few officials, though the effectiveness of the system depended on the strength of indigenous institutions. European settlement in countries such as Kenya and Zimbabwe (then Southern Rhodesia) was another feature of British colonial policy, much of the best land being taken over by the settlers, whilst African farmers were crowded into less fertile reserves, often with disastrous ecological results. In the case of Zimbabwe, the persistence of white settlement and white minority rule led to a considerable delay in the granting of independence. In British West Africa there was no significant white settlement and the ultimate granting of independence to the four territories, Nigeria, Ghana (then Gold Coast), Sierra Leone and The Gambia, was much less troublesome.

The French approach to her colonial territories was very different from the British approach. Whereas British colonialism was designed to create Africans with British characteristics, French policy was designed to create black Frenchmen. French colonialism was more direct and centralized, the colonies being regarded as part of France and referred to as 'Overseas France'. French policies were essentially geared towards assimilating the colonies into France and the French way of life. Large numbers of French administrators went into the African colonies and, unlike the British, there was little attempt to work through traditional institutions or leaders. Local cultures and languages were given little encouragement. The French believed in a bureaucratic and hierarchical state and created small indigenous elites in their territories whose members were often educated in France. White settlement was particularly common in the African countries bordering the Mediterranean, but there were also sizeable white minorities in tropical Africa, such as in Dakar (capital of Senegal) and in the plantation areas of Cameroon, Guinea, Ivory Coast and Madagascar.

Belgium controlled one vast territory, the Belgian Congo (now Zaire), initially as the personal possession of King Leopold. The Belgians adopted a policy of assimilation similar to the French, but like the British policy its ultimate aim was independence. Emphasis was placed on economic and social development with a focus on elementary education, hence there was no significant growth of an indigenous educated elite. The Belgians argued that Africans needed guidance and tuition before they could take over, and political expression was suppressed until just before independence.

Portugal was the other major colonial power, a country which played a key role in making initial contact with the African coast during the voyages of discovery in the fifteenth century. As early as 1482 the Portuguese established fortified trading posts on the coast from Senegal to the Gold Coast and were responsible for naming the territory which became Sierra Leone. In the scramble for Africa, however, Portugal was less successful in gaining territory than either Britain or France. By the early twentieth century Portugal controlled Portuguese Guinea (now Guinea-Bissau) in West Africa, and the two much larger colonies of Angola and Mozambique. Portugese policy, like the French, was aimed at assimilation, with all territories being regarded as part of the Portuguese Union. There were also substantial numbers of white settlers in the Portuguese colonies.

The other colonial powers in tropical Africa were Spain, which controlled Western Sahara and Spanish Guinea (now Equatorial Guinea), and Germany which controlled Togo and Cameroon in West Africa, and Tanganyika and Ruanda Urundi in East Africa. However, Germany's hold on African territory was short-lived as after the First World War her colonies were mandated to other European powers, Britain, France and Belgium.

One significant and very unpopular element in the establishment of administrative frameworks in the African colonies was the introduction of monetary taxes, to be paid at all levels and at particular times. For many Africans taxation of any kind was a complete innovation, whilst others had only paid indirect taxes. In order to obtain money to pay these taxes, cash crops had to be grown. Additionally, many people, usually men, had no alternative but to leave their villages and go to the rapidly growing towns and plantations where wage labour was needed. There are many instances of African people demonstrating their displeasure at colonial taxation, but tax evasion was brutally discouraged and could lead to harsh punishment and forced labour. In fact, the imposition of colonial rule in general was often harsh and any form of resistance to the colonial state was dealt with severely.

The economic policies pursued by the colonial powers had a profound effect on the African people. The widespread cultivation of cash crops for export was one such policy. In French West Africa and in northern Nigeria the groundnut became the principal cash crop, grown primarily to provide oil for growing industries in France and Britain. The cultivation of cash crops often meant that food crops were neglected and the nutrition of rural people suffered. The situation became particularly severe during drought periods when many people died of starvation. In other parts of tropical Africa cash crops were encouraged, not so much to meet demands in Europe, as to find some practical form of revenues. In Kenya, for example, sisal, coffee and maize made the colony self-supporting.

As each colony was viewed not as a separate economic unit, but as a cog in the imperial machine, there was a tendency towards specialization in some countries and regions. This was the origin of the 'one-product economy', which is still a feature of many African countries today; for example, copper from Zambia, groundnuts from The Gambia and Senegal, cocoa from Ghana and diamonds from Sierra Leone. As a result of such specialization, these and other countries were, and still remain, vulnerable to world price fluctuations and dependent on the colonial powers and industrial countries for markets, capital and technology. There was little trade between colonies and any profits went to the colonial power rather than being invested in the colony.

The period leading up to independence was often characterized by a lack of preparation for the event and its responsibilities. In the French and British territories the preparation for taking over the legislative, executive and administrative roles and structures by Africans began only after the Second World War. Few chiefs in British Africa were allowed any initiative in the administration before 1945. In Guinea, the French actually tried to destroy the fabric of the state before they departed, even removing the books from the law library of the Ministry of Justice. However, the Portuguese territories and the Belgian Congo probably provide the most notorious examples of the lack of preparation for the transfer of the institutions of state.

A question that is often asked, is to what extent can colonialism be blamed for Africa's current problems thirty or more years after independence? Indeed, the colonial experience and its aftermath remain major topics of debate amongst historians and there are many conflicting viewpoints. Suffice it to say here that colonialism initiated a type and pace of change that was unprecedented in Africa. Colonial policy was generally exploitative and often coercive and confrontational. Although it is important to evaluate the many other factors which may have contributed to the present condition of African countries, such as political instability and economic mismanagement, it would be inappropriate to ignore the profound effects of colonial policies and their legacy.

Problems, challenges and prospects

Most African countries have a long way to go if they are to achieve the living standards and political and economic stability enjoyed by countries in Western Europe and North America. Some of tropical Africa's 43 states are among the poorest in the world. Mali, in the Sahel region of West Africa, for example, is more than five times the size of the United Kingdom, but has a population of only 8.2 million compared with a population of 57.4 million in the UK. Mali had a per capita Gross National Product (GNP) in 1989 of only 270 US dollars ($US), compared with a figure of $16,100 in the UK. Life expectancy at birth in Mali is 48 years compared with 76 years in the UK, and in 1988 Mali had a daily per capita calorie supply of only 2181, compared with the UK figure of 3149. Mali's situation, like that of so many other African countries, could deteriorate further in the next decade, with a relatively high projected population growth rate of 3.0 per cent per annum compared with only 0.2 per cent in the UK. Furthermore, Mali is one of Africa's 14 landlocked states and is situated in a zone where rainfall is notoriously unreliable and drought a regular occurrence.

But not all African countries are like Mali, and it is important to be aware of the great diversity. Within mainland tropical Africa, Gabon had the highest per capita GNP of $2960 in 1989. Such national statistics, however, conceal major variations within individual countries. It is common in Africa for people in urban areas to be relatively better off than those living in rural areas. In fact the differential between the 'haves' and the 'have-nots' in African countries is far greater and frequently more visible than in most developed countries. Fleets of air-conditioned Mercedes Benz cars, sumptuous residences guarded with high gates, electrified fences and barbed wire, modern hotels and banks are now common features of most African capitals. There is poverty in African cities too, often in close proximity to the symbols of wealth. Wooden, or even cardboard shacks, with no running water supply, proper sanitation or electricity can be found in many large African cities, and it is likely that there will be more of them in the future.

African countries face many problems and challenges. Most have suffered greatly since the early 1970s with the knock-on effects of the oil price rise in 1974–5 and the subsequent world recession. Mounting debt and frequent shortages of basic commodities due to a lack of foreign exchange are commonplace. Non-oil producers experience regular oil and petrol shortages, preventing the movement of goods and people and further compounding already serious economic problems. In cities, electric power cuts are a regular occurrence, caused by fuel shortages or inadequate maintenance of ageing generating plants and power lines. The effects of prolonged power cuts on industry can be devastating.

It is in the rural areas, however, that most of Africa's population live and work. Some would argue that the towns have already received too much of the wealth and investment, and that there is now a need for a

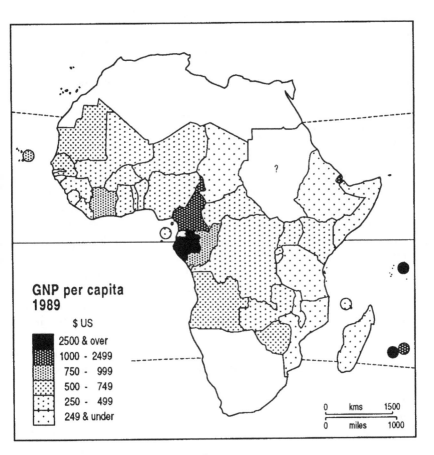

Figure 1.3 GNP per capita ($US 1989)

genuine and concerted effort to raise rural living standards and make agriculture more attractive and profitable. In most tropical African countries, over 60 per cent of the labour force works in agriculture, mainly producing traditional food crops and using methods which have been handed down from generation to generation with little change. Some countries have a very high proportion of their labour force engaged in farming, such as Rwanda and Burundi, Niger, Burkina Faso, Mali and Tanzania, where in all cases over 80 per cent of the labour force is in agriculture. Although the proportion working in agriculture is declining slowly, it is likely to remain high for the forseeable future, and contrasts sharply with a figure of under 3 per cent employed in UK agriculture. For those rural young people with some education, agriculture is much less attractive than the prospects of urban employment. African governments need to restore faith in agriculture by helping farmers to raise production and by paying them good prices for their crops. Greater food production should reduce the need for food imports which, together with possible food exports, would improve the foreign exchange position. As farming systems become more efficient, so they may gradually shed surplus labour to the growing industrial sector, usually located in or near to the towns. Improving Africa's agriculture will take time and must be done carefully. Above all there is a need to take account of the many failures in the past, and for governments and agricultural advisers to work together with farmers rather than merely instructing them which crops they must grow and which techniques they should use. African leaders are fond of talking about the importance of agriculture, but very often little is done, their concern being to keep food prices low in the towns so that the better organized, and politically more threatening, urban population stays contented.

A further priority for the future will be to improve the transport systems of African countries, which in many cases were laid down in the colonial period. Roads with crumbling surfaces and pot-holes in the rainy season, and railways with worn-out track and rolling stock are all too common. If agriculture is to become more productive and the benefits of this more widely distributed, then an efficient marketing and transport system is absolutely vital. Much investment is needed, yet many African governments cannot afford the outlay.

Africa's future is both exciting and uncertain. There is great potential, but the means, predominantly financial, of achieving this potential are sadly lacking. Many African countries are now relatively worse off than they were at independence more than thirty years ago. World recession, indebtedness and shortage of foreign exchange, coupled with limited achievements in agricultural development, poor health and welfare, rapid population growth and crumbling infrastructures, have all contributed to the poverty in many African countries today. Furthermore, economic instability has often bred political instability and vice versa. One-party states, military regimes and *coups d'état* are common features in Africa, making potential foreign investors cautious about investing in unpredictable locations. The two great capitalist 'success stories' in Africa, Ivory Coast and Kenya, have enjoyed relatively long periods of political stability, suggesting that perhaps political and economic stability go hand in hand. Many other African countries, such as Chad,

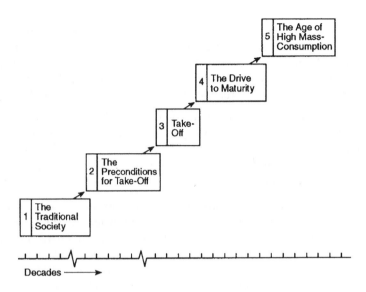

Figure 1.4 The Rostow model

Ethiopia, Mozambique, Somalia and Uganda, have suffered long periods of political instability and even internal wars. Economic stability and progress are more likely to be fostered if African countries could settle their internal differences of opinion more calmly and generate an atmosphere of greater political stability. Possibly only then will really significant advances in economic and social development be possible.

Perspectives on tropical Africa's future development

The 1970s and 1980s have been a particularly difficult time for tropical Africa, with some countries actually experiencing negative economic growth rates. Mozambique, for example, had an average annual growth rate of −0.7 per cent between 1980 and 1990, whilst Niger's rate was −1.3 per cent, amongst the lowest rates in the world. At the other end of the spectrum in tropical Africa, Congo Republic had an annual growth rate of 3.6 per cent between 1980–90 and Kenya 4.2 per cent.

In 1960, W. W. Rostow, in his 'stages of economic growth' model (see Figure 1.4), suggested that economic progress in particular countries could be categorized into certain phases, of which the most important was 'take off', a period associated with increased industrial investment such as occurred during the British industrial revolution in the late-eighteenth and early-nineteenth centuries. The Rostow model seems to imply that all countries proceed through the same five phases of growth, that industrialization is the key to growth and that growth leads to economic and perhaps social improvement. The model has, however, been heavily criticized in recent years on the following grounds:

- its unilinear approach, implying that 'things get better' over time;
- its Eurocentricity, in implying that all countries will imitate the Euro-American experience;
- the confusion generated concerning the definition and character of growth and development.

Some writers have drawn attention to the existence in some developing countries of 'growth without development' where, despite high rates of economic growth (as measured by increases in Gross Domestic Product), the bulk of the population does not seem to have benefited as one might have expected. Dudley Seers draws a clear distinction between development and economic growth. Development, Seers suggests, should involve a reduction in poverty, unemployment and inequality, whilst individuals and groups should also be able to fulfil their ambitions. In many cases where economic growth has occurred, only a small minority of the, usually urban-based, population has benefited. This is what happened in Liberia in the 1960s when the small Americo-Liberian ruling group benefited from the country's rapid economic growth largely based on mining and rubber production. The benefits of such growth should ideally be fairly distributed and penetrate into the remotest areas of countries, rather than being 'siphoned off' by a relatively small number of people located at the nation's political and economic hub. But how can this be achieved, and how can any progress be monitored effectively?

In most developing countries it would be difficult to obtain reliable statistics on poverty, inequality and unemployment. Furthermore, it would be necessary to define each of these terms carefully in relation to the situation being examined. For example, poverty in Britain may be very different in character and magnitude from poverty in Burkina Faso or Zambia. Whilst in Britain reference would probably be made to minimum wage levels and receipt of supplementary benefit or other allowances, in many African countries similar social security measures do not exist. In an African context, some reference to nutritional and health status as well as income would perhaps be more meaningful.

Similarly, measuring inequality and unemployment in most African countries would be difficult without reliable census and other socio-economic data. Registers of unemployed people are usually only kept, if at all, in African cities, and even there they are unlikely to reflect the situation accurately. Employment is more seasonal in rural areas, being related to cycles of cultivation, with the heaviest work coming during the rainy season. The dry season may be a period of un- or rather under-employment, when other jobs such as house-building or craft work are undertaken in rural villages. Alternatively, for those persons living close to large towns, there may be an opportunity to take up seasonal paid work in the town, perhaps as a carpenter, mechanic or in some other trade. Such off-farm dry season work can make a significant contribution to household income as in the region around the city of Kano in northern Nigeria.

Whilst we might agree with Seers that definitions of development should include reference to a series of non-economic as well as economic variables or indicators, it is often difficult to measure these in many African states. Seers suggests that once the magnitude of poverty, unemployment and inequality is recognized, governments and development agencies should introduce programmes to fulfil basic human needs and reduce these three key indicators of development. The poorest groups should be the main focus of development planning, with particular attention directed towards provision of adequate food, water, shelter, health care, education and employment – the basic needs of all human beings.

It has taken a long time for African governments to take on board these ideas, and some remain committed to the much-criticized Rostow model. There is, however, an increasing move towards smaller-scale, village-based projects aimed at providing basic needs to the poorest groups. Whereas earlier projects and programmes for promoting economic growth usually focused on wealth generation, more recent development projects in tropical Africa are increasingly focused on poverty alleviation. It is argued that if basic needs are fulfilled people will be able to perform more efficiently and live longer, more fruitful lives, which in turn will raise their economic output in the long term. Thus, fulfilling basic needs will eventually lead to economic growth as well as development. In later chapters of this book reference will be made in greater detail to specific examples of development schemes and programmes. Although many African countries are starting to follow a basic-needs approach to development, they still have a long way to go, and it is likely that widespread poverty, inequality and unemployment will remain for some time.

Case study A

The Gambia – colonial legacy

With an area of only 11,300 sq. km (smaller than Northern Ireland), and a population in 1989 of 848,000, the West African state of The Gambia is both one of tropical Africa's smallest countries and one of the poorest. Its per capita gross national product of about $240 (UK $14,570), according to the World Bank, makes it the fifteenth poorest country in the world.

It is something of a miracle that The Gambia has managed to survive as a state in its own right for over a century. It is a long, thin country, 350 km from west to east and no more than 48 km wide, at the heart of which is the River Gambia, the focus of life and economic activity. Until independence in 1965, The Gambia was a British colony and the official language is English. The administrative framework, together with the health, education and judicial systems, were put in place by the British during more than a century of colonial rule. Banjul (formerly Bathurst), the capital city and once the hub of the colonial state, is now a rather sleepy, declining settlement of around 60,000 people. Many of the features of a colonial town are still evident in its layout and the style of buildings, and government ministries, foreign consulates and development agencies are located there.

The Gambia is surrounded on three sides by the French-speaking state of Senegal, which is more than seventeen times larger. The Gambia has been described as 'a hot dog in a Senegalese roll', or as Leopold Senghor, former President of Senegal, said in 1961, The Gambia is '. . . pointed like a revolver in the very stomach of Senegal'. Senegal's administrative structure is descended from that introduced by the French colonial rulers before independence in 1960. Although the bureaucracies of The Gambia and Senegal are very different, the social and cultural features of their peoples show strong similarities and a common disregard for the artificial international border which cuts through their homelands. Most of The Gambia's people are Muslim, with

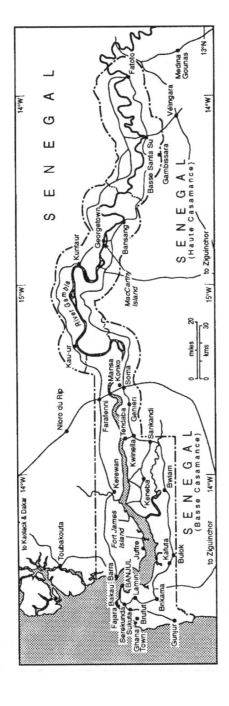

Figure 1.5 The Gambia

Case study A (*continued*)

the Mandinka and Wolof comprising the main cultural groups. In Senegal, the same cultural groups are also important and there is constant movement of people and goods across the border dividing the two countries. Many families in The Gambia have relatives in Senegal and visit them regularly. Farm workers from Senegal have for many years migrated into The Gambia to assist with the harvest. At Mansa Konko, 150 km up the River Gambia, the important Trans-Gambia Highway crosses the country, linking northern Senegal and Dakar, its capital, with the southern region of Senegal, known as Casamance. With an international border over 1,000 km it has, understandably, always been difficult to control the movement of people and goods between the two countries and smuggling has been a constant drain on the two economies.

For a short period in the 1980s it did seem that The Gambia and Senegal might be coming closer together in the Confederation of Senegambia. The confederation was a direct result of a Libyan-backed coup in The Gambia on 30 July 1981, whilst the country's president, Sir Dawda Jawara, was attending the wedding of Prince Charles and Lady Diana Spencer in London. During the brief uprising, 500 people were killed and over £10 million worth of damage was done, mainly in and around Banjul. In 1981 The Gambia had no national army and to everyone's surprise, Sir Dawda invited Senegalese troops to enter his country and put down the rebellion. Sir Dawda returned to his liberated country on 2 August, 1981 and within six months he and the president of Senegal, Abdou Diouf, proclaimed the Confederation of Senegambia. Some progress was made on trade agreements and the establishment of monetary and customs union. However, in August 1989 the confederation collapsed when Senegal pulled out after Jawara suggested that the confederation presidency should rotate between Diouf and himself. After the failure of this brief and unexpected experiment, there is today a widespread feeling that in the end both countries need each other and it remains to be seen if a further period of cooperation develops.

One of the legacies of the colonial period is that The Gambia is a

Case study A (*continued*)

Plate 1.1 A rural skills workshop sponsored by a non-governmental organization in The Gambia. School leavers learn how to repair tools and do simple metalwork.

one-product economy dominated by groundnuts. Groundnut cultivation was encouraged by the British, the crop being first exported in the 1830s. A striking feature of many Gambian villages is that women comprise over 70 per cent of the agricultural labour force and produce more than 80 per cent of domestic food requirements. When groundnuts were introduced it was the men who mainly adopted the new cash crop, leaving the women to grow rice, millet, maize, sorghum and a large variety of vegetables. Cash obtained from groundnut sales is a key element in most families' income, and any change in the price paid for groundnuts can have widespread repercussions in rural communities. In the late 1980s groundnuts accounted for 80 per cent of domestic exports, about 60 per cent of total crop land and contributed around 35 per cent

Case study A (*continued*)

of the country's gross domestic product (GDP), though this figure has gradually declined in recent years due to the increasing importance of manufacturing, commerce and particularly tourism.

The Gambia provides an ideal location for the development of tourism, catering for Europeans who want to exchange the cold and gloom of winter for guaranteed sunshine and temperatures around thirty degrees centigrade. Some 112,800 tourists visited the country in the 1988–9 tourist season. With an estimated average expenditure per tourist of $US 400, this amounted to about $US 45 million from the tourist industry, representing 10 per cent or more of GDP. In recent years over half of all tourists came from Britain and the remainder from Scandinavia (notably Sweden), France and Germany. The present and potential contribution of tourism to Gambian development must be considered carefully. Already it is estimated that some 7,000 people are working in hotels and tourist-related activities. But there is a danger that in such a small country tourism could easily swamp the local economy and society and cause an increase in theft, begging and prostitution.

Tourism at present only affects a small proportion of Gambia's land and people, and for over 75 per cent of the population traditional forms of agriculture and pastoralism provide their main livelihood. In much of rural Gambia methods of food and livestock production have probably changed very little over hundreds of years. Relative to the tourists, many of the local people are very poor indeed. Life expectancy in The Gambia is only 44 years, compared with 76 years in the UK, and some 75 per cent of the population is illiterate. The statistic which perhaps best highlights the level of poverty is infant mortality, which in The Gambia is 139 per 1000, compared with 9 per 1000 in the UK. With a population growth rate of around 3.3 per cent per annum, and a density of over 75 people per square kilometre, The Gambia is already one of the most densely settled countries in West Africa, and twice as densely settled as neighbouring Senegal.

Clearly, The Gambia does have major problems. Its size, shape and the dominance of groundnuts in the economy are a legacy of the colonial period. Whilst tourism has helped to diversify the

Case study A (*continued*)

economy and provide much-needed income and jobs, for the bulk
of the country's people there has been little improvement in rural
living standards since independence.

Key ideas

1 Longstanding myths and stereotypical views of tropical Africa are
   difficult to eradicate and inhibit accurate understanding.
2 Although the colonial period was a relatively short phase in Africa's
   history, the effects of colonialism have been far-reaching.
3 Inadequate understanding of Africa's peoples and environments has
   often led to inappropriate development strategies and projects which
   have sometimes caused more problems than they have solved.
4 Considerable social and economic inequalities exist between and
   within tropical African countries.
5 Rates of economic growth have been negative in some African
   countries in recent years. In countries where positive economic
   growth has been achieved, development and the fulfilment of basic
   needs have often not occurred.

2
Africa's people

Introduction

In the past, Africa has been a continent short of people rather than land. The presence of domestic slavery long before the colonial powers arrived, and continuing in some places until independence, testifies to the importance of owning labour. However, in a study of Africa's population, there is as much variation across the continent today as there was in the past. The homelands of the Ibo and Yoruba peoples in southern Nigeria, the highlands of Ethiopia and the two tiny states of Rwanda and Burundi, west of Lake Victoria, have long been densely populated regions. On the contrary, countries such as Niger and Chad, bordering the Sahara Desert, together with the Central African Republic and the Zaire (Congo) Basin have always been sparsely settled. But tropical Africa's population is increasing rapidly and pressure on land and other resources is growing in some cases to dangerous levels.

Counting tropical Africa's people

The total population of the 43 states which, for the purpose of this book, comprise tropical Africa, is approaching 476 million. But this figure can only be approximate, since much of the data relating to the African population is unreliable for a number of reasons. Censuses are expensive and require skilled personnel to administer them, and most African countries are desperately short of finance and expertise. In many countries a large proportion of people live in inaccessible rural areas and are illiterate, so need help in completing their census forms. They are often suspicious of censuses, associating them with taxes and law enforcement. For these and other reasons most African countries lack detailed population data and series of data going back several decades are almost totally lacking.

In most countries, the first censuses were conducted after the Second World War by the colonial powers: amongst the British colonies, Uganda, Kenya, Tanganyika and the Gold Coast were the first to have full censuses in 1948. Sierra Leone had to wait until 1963 for its first full census, whilst Cameroon did not have a census until April 1976.

Conducting a census can be a sensitive issue. The results of the December 1974 census in Sierra Leone were witheld for many years for fear of disturbances that might be caused if one tribal group was seen to have more members than another. Perhaps the most extreme case is that of Nigeria where, until 1991, the 1952 census was the last one widely accepted as reasonably accurate for the country as a whole. The 1962 and 1963 censuses had totals which differed by more than 10 million persons and the results of the 1973 census were abandoned for the same reason. Although Nigeria is widely regarded as the most populous state in Africa, the exact size of its population has not been known until recently. Various estimates were put forward in the late 1980s, in some cases suggesting the population was already well over 100 million. For example, the 1989 estimate given in Table 2.1 was 113.6 million. President Babangida, Nigeria's military leader, declared that a full national census would be taken in 1991, prior to a return to civilian rule in 1993. The provisional results of the November 1991 census revealed a total population of 88.5 million, considerably lower than most forecasts.

Censuses must be a priority in tropical Africa. Regular and reliable population data are vital if countries are to plan for their future development and for the building of schools, hospitals, roads, water and sewage facilities. But with such inadequate resources, a census may seem something of a luxury when poverty and malnutrition are widespread and needing urgent attention. Many governments have to be content with sample surveys which are cheaper and easier to conduct, but unfortunately do not give a complete picture.

Table 2.1 Tropical Africa: basic statistics

	Area (sq. km)	Population (thousands), 1989	Population density (persons per sq. km)	Average annual population growth rate, 1980–89, percentage	Life expectancy at birth (years), 1989	Infant mortality rate (aged under 1) per 1,000 live births, 1989	Daily calorie supply per capita, 1988	Gross National Product (GNP) per capita, US$1989
Angola	1,246,700	9,694	7.8	2.5	45	132	1,725	610
Benin	112,622	4,593	40.8	3.2	51	112	2,145	380
Burkina Faso	274,200	8,776	32.0	2.6	48	135	2,061	320
Burundi	27,834	5,299	190.4	2.9	49	70	2,253	220
Cameroon	475,442	11,554	24.3	3.2	57	90	2,161	1,000
Cape Verde	4,033	369	91.5	2.5	66	*42	2,436	780
Central African Republic	622,984	2,951	4.7	2.7	51	100	1,980	390
Chad	1,284,000	5,537	4.3	2.4	46	127	1,852	190
Comoros	2,400	459	191.3	3.7	57	*95	2,046	460
Congo	342,000	2,208	6.5	3.4	54	115	2,512	940
Djibouti	22,733	410	18.0	3.5	48	*118	⋮	*1,210
Equatorial Guinea	28,051	344	12.3	2.0	46	*123	⋮	330
Ethiopia	1,221,900	48,861	40.0	2.9	48	133	1,658	120
Gabon	267,667	1,105	4.1	3.7	53	98	2,396	2960
Gambia	11,295	848	75.1	3.3	44	*139	2,360	240
Ghana	238,537	14,425	60.5	3.4	54	86	2,209	390
Guinea	254,857	5,547	21.8	2.5	43	140	2,042	430
Guinea-Bissau	36,125	960	26.6	1.9	40	*141	2,690	180
Ivory Coast	322,463	11,713	36.3	4.0	53	92	2,365	790
Kenya	582,645	23,277	40.0	3.8	59	68	1,973	360
Liberia	111,369	2,475	22.2	3.1	50	137	2,270	*450
Madagascar	687,041	11,174	16.3	2.8	51	117	2,101	230
Malawi	118,484	8,230	69.5	3.4	47	147	2,009	180
Mali	1,240,000	8,212	6.6	2.5	48	167	2,181	270
Mauritania	1,030,700	1,954	1.9	2.6	46	123	2,528	500

Mauritius	2,045	1,062	519.3	1.0	67	*22	2,679	1,990
Mozambique	783,030	15,357	19.6	2.7	49	137	1,632	80
Niger	1,267,000	7,479	5.9	3.5	45	130	2,340	290
Nigeria	923,768	113,665	123.0	3.3	51	100	2,039	250
Reunion	2,510	585	233.1	1.6	72	*13	2,665	*5,570
Rwanda	26,338	6,893	261.7	3.3	49	118	1,786	320
Sao Tome & Principe	964	122	126.6	3.0	66	*45	2,657	340
Senegal	196,192	7,211	36.8	3.0	48	82	1,989	650
Seychelles	518	68	131.3	0.9	70	*18	2,146	4,230
Sierra Leone	71,740	4,040	56.3	2.4	42	149	1,806	220
Somalia	637,657	6,089	9.5	3.0	48	128	1,736	170
Sudan	2,505,813	24,423	9.7	3.0	50	104	1,996	
Tanzania	945,087	25,627	21.1	3.5	54	112	2,151	130
Togo	56,000	3,507	62.6	3.5	54	90	2,133	390
Uganda	236,036	16,772	71.1	3.2	49	99	2,013	250
Zaire	2,354,409	34,442	14.6	3.1	53	94	2,034	260
Zambia	752,614	7,837	10.4	3.7	54	76	2,026	390
Zimbabwe	390,580	9,567	24.5	3.6	63	46	2,232	650
TOTAL/AVERAGE	21,718,383	475,721	64.8	3.0	52	103	2,147	717

Sources:

1 Columns 1, 2, 4, 5, 7: *World Bank Atlas*, 1990, Washington, World Bank.

2 Column 6: *World Development Report*, 1991, Oxford, Oxford University Press, except those marked *, from *Basic Statistical Data on Selected Countries*, 1991, London, Commonwealth Secretariat.

3 Column 8: as for column 6, except those marked *, from 1987 figures from Commonwealth Secretariat, 1991.

Population size, distribution and density

By far the most populous state in tropical Africa is Nigeria with a population of 88.5 million, almost twice as many people as the next most populous country, Ethiopia, which has a population of 48.8 million. At the other end of the spectrum, some of the island states have the smallest populations. Seychelles, for example, has only 68,000 people, whilst São Tomé and Principe have a total population of 122,000. Of the mainland countries, those with the smallest populations are Djibouti (410,000), Equatorial Guinea (344,000), The Gambia (848,000), Guinea-Bissau (960,000) and Gabon (1,105,000).

Most of tropical Africa's 476 million people live in rural areas, mainly in scattered villages. By world standards it is not a densely settled region. The average density of 64.8 persons per square kilometre includes small, heavily populated island states such as Mauritius (519.3 per sq. km) and Reunion (233.1 per sq. km). By contrast Mauritania, on the southern edge of the Sahara, has a density of only 1.9 per sq. km and Gabon, astride the equator, 4.1 per sq. km. These figures may be compared with the population densities of other countries such as the UK with 233 per sq. km, India 253, and the Netherlands 400.

Even within particular countries there are densely settled and sparsely settled areas. The coastal region of West Africa from Ivory Coast east to Cameroon, the Nile valley and the area to the north and west of Lake Victoria are the most densely settled regions on the mainland, together with some smaller, but equally dense, pockets such as around Kano in northern Nigeria. Sparsely settled areas include the Sahara Desert and the countries on its southern fringe, the eastern horn, Central African Republic and the Zaire (Congo) basin.

There are many reasons why some areas are densely settled whilst others have very few people. Other than along the Nile, river valleys have not been popular places for settlement in tropical Africa, owing to the presence of water-borne diseases such as bilharzia (schistosomiasis), river blindness (onchocerciasis), malaria and others. People have avoided some areas because of harsh climate, notably the lack of water, as in many countries close to the Sahara Desert. In West Africa, particularly Nigeria, a region known as the 'middle belt' exists in between densely settled areas in the north and south. Poor soils and the widespread presence of the tsetse fly are problems here, but the sparseness of population is probably due mainly to slave raiding in the eighteenth and nineteenth centuries before the onset of colonial rule.

Plate 2.1 The village of Fanisau north of Kano, Nigeria, showing the structure of family compounds, with, in the distance, the intensively cultivated farmland of the Kano close-settled zone.

As people were cleared from such areas, the bush grew back over farmland and the tsetse fly became a greater problem, spreading sleeping sickness amongst the remaining inhabitants. Densely settled areas are usually a result of complex environmental and social reasons. Upland areas such as the Ethiopian highlands and the Jos Plateau in Nigeria were popular as refuges during times of inter-tribal warfare, and inhabitants have developed intensive systems of terrace farming in difficult environments (Case study E). Fertile volcanic soils have attracted dense settlement in West Cameroon, Rwanda and Burundi. The high quality brown soils of the Kano close-settled zone in northern Nigeria support one of the densest rural populations in tropical Africa. Other areas of dense settlement include mining centres such as the Shaba-Copper Belt region of Zaire and Zambia.

Major urban centres have also attracted population, although in many countries fewer than 25 per cent of the total population live in towns. In Burundi for example, the figure is only 6 per cent, Burkina Faso 9 per cent and in Ethiopia 13 per cent. In West Africa many of the largest

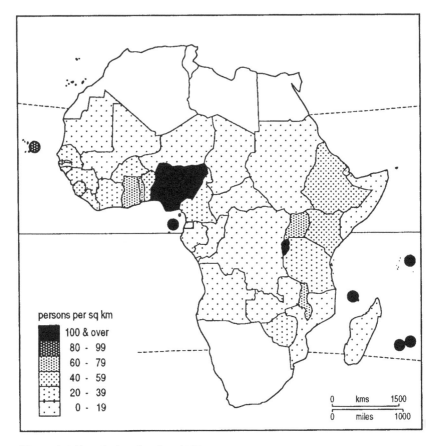

Figure 2.1 Population density, 1989

settlements are on the coast and have become a focus for dense population. Away from the coast, the regions surrounding Nairobi in Kenya and Kampala and Entebbe in Uganda have attracted dense population.

Population growth

Evidence suggests that the total population of Africa was fairly stable at around 100 million for many years, probably from the mid-eighteenth century to the late-nineteenth century when the colonial powers arrived. It is estimated that in 1900 Africa's population was about 133 million and that this more than doubled in sixty years to reach 270 million in

1960 on the eve of independence. By 1970 the population had reached 344 million and in the mid-1980s (1985), the figure was approaching 550 million for the continent as a whole and 410 million for those countries in tropical Africa.

The period since the Second World War (post-1945) has seen a rapid acceleration in the growth of Africa's population, such that today some countries have population growth rates which are among the highest in the world. Growth rates above 3.5 per cent per annum are not uncommon, as in Ivory Coast (4.0 per cent) and Kenya (3.8 per cent). Everywhere in tropical Africa population is increasing, unlike much of Western Europe where growth is minimal or non-existent. Some African countries, however, do have lower population growth rates, notably island republics such as Reunion (1.6 per cent), Seychelles (0.9 per cent) and Mauritius (1.0 per cent). On the mainland however, only Guinea-Bissau (1.9 per cent) has a growth rate of less than 2 per cent per annum.

Demographic structure

Since 1945 most African countries have moved into the second stage of the so-called 'demographic transition', lagging behind many other parts of the world. From a situation of high birth rates and high death rates (stage 1), death rates have been reduced in recent years through medical improvements, whilst birth rates have remained high. Only in a few African countries have birth rates begun to fall. In Mauritius, for example, a vigorous family planning programme helped to reduce the birth rate from around 47.4 per thousand in the years 1949–53 to 26.7 in 1970. By 1985 the figure had fallen even further to 20 per thousand. For many other African countries, however, there was little movement in the birth rate. The net effect of lower death rates combined with high birth rates has been a rapid growth of population.

Demographic structure is best seen from a population pyramid which shows the proportions of males and females in different age groups. A broad base indicates a high birth rate, whilst a narrow peak shows a high death rate and low life expectancy. In many African countries the population pyramid has a broad base with a large proportion of the total population in the lower age groups. The pyramid tapers, with relatively few people in the upper age groups. In most countries life expectancy is well below the seventy or more years that people in Western Europe and America can expect to live. In fact, in much of tropical Africa the majority of people can only expect to live fifty years or even less. For example, life expectancy at birth is only forty years in Guinea-Bissau. In some crowded island states, however, people can expect to live much longer. Life expectancy in Mauritius, Seychelles and Reunion, for example, is sixty-seven, seventy and seventy-two years respectively. However, these three countries have managed to reduce their birth rates significantly, resulting in lower than average population growth rates.

It is interesting to compare the population pyramids of two countries such as Mauritius and Zaire (see Figure 2.2). Mauritius, with a low birth rate, has a narrower base to its pyramid and a small proporton of its population in the lower age groups. With a high life expectancy there is a good proportion of both males and females above the age of fifty. The population pyramid for Zaire, however, has a broad base, reflecting a high birth rate, and a greater proportion of the population in the lower age groups. The peak of the pyramid is very sharp, with relatively few people living over fifty years.

Redistribution of population

Apart from natural increase in population, the other main factor affecting the size and distribution of population is redistribution, mainly through population movements of one sort or another. Movements of people within and between the countries of Africa have in some cases been going on for centuries. We can distinguish between two main types of movement; migration and circulation. Migrations are usually permanent or irregular, but involve a lengthy change of residence, often longer than a year. Circulations are usually shorter, sometimes daily, periodic or seasonal, or in some cases long term.

There are many examples of migrations in tropical Africa, some are voluntary, whilst others are enforced. In West Africa, the movement of Mossi men from Burkina Faso to help with the cocoa harvest in Ghana and Ivory Coast has been going on for many years and the money they earn is a vital addition to their poor villages in Burkina. Others migrate to towns for education or to attend a hospital, whilst a large number of young people are attracted by the 'bright lights' and relatively modern facilities to be found in towns, particularly capital cities. Economic reasons are the main cause of rural–urban migration in most of Africa. For many, a spell in a big city is regarded as an initiation into adulthood and western culture. In central and southern Africa, an important

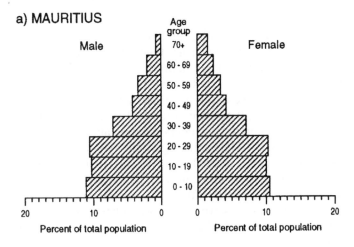

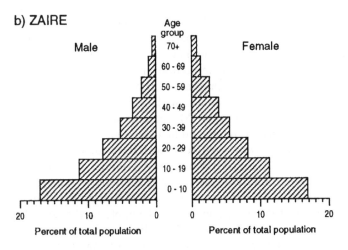

Figure 2.2 Population pyramids of Mauritius and Zaire, 1989

migration involves the movement of thousands of people, mainly young males, to the industrial areas of South Africa. At any one time a large number of people from Zambia, Zimbabwe, Malawi and Mozambique are absent from their homes and all agricultural and domestic work has to be done by the women who remain behind.

Tropical Africa also has many examples of refugees who have

migrated to avoid civil war and persecution in their homelands. Angola, Mozambique, Somalia, Rwanda, southern Sudan and Ethiopia are just some of the many areas from which refugees have migrated in search of sanctuary. In the 1980s, large numbers of refugees crossed from Ethiopia into eastern Sudan and from Chad and Uganda into western and southern Sudan. Refugees have also abandoned their homes in Mozambique, escaping the violent attacks of RENAMO (Mozambique National Resistance) on the civilian population and the general state of famine in that country. Malawi has been the main destination of these refugees. More recently, in 1990 and 1991, the civil war in Liberia has caused many refugees to cross into neighbouring countries such as Sierra Leone with devastating effects on economy and society, particularly in the country's Eastern Province. Other enforced migrations include the repatriation of Ghanaians from Nigeria and Mauritanians from Senegal.

One longstanding, long-distance migration involves the movement of pilgrims to the holy city of Mecca in Saudi Arabia. All Muslims aim to make the Hadj at least once during their lives. Whereas this migration was once done by foot taking many months, today the round trip to Mecca can be done in a few hours by air. At Kano airport in northern Nigeria, for example, a whole section of the airport is devoted to the movement of thousands of pilgrims each year.

Various types of circulation usually take place over a shorter distance and time-span than migration. Movements between rural and urban areas are particularly common and involve large numbers of people. On a daily basis, the journey to work in town from surrounding rural areas is widespread. In many African countries public transport is at best inadequate, and often non-existent. It is common to see people travelling to work in taxis, converted trucks and minibuses, on bicycles or, most often, on foot. Large numbers of people in the so-called 'peri-urban' areas are dependent for their livelihood on urban employment.

Others make the regular journey into town to sell produce grown in market gardens and farms. In Sierra Leone, for example, villages on the mountain slopes behind the capital, Freetown, have intensive market gardening of fruit and vegetables. Each morning before dawn, women carry heavy baskets of produce downhill to markets in the capital. Others hire taxis or trucks to carry foodstuffs. Around Kano in northern Nigeria, a similar intensive production of vegetables can be seen, in this case making use of low-lying depressions known as *fadamas*, which in many cases remain cultivable throughout the long dry season. Similarly,

Plate 2.2 Women selling cassava leaf at a rural market in Sierra Leone.

close to Banjul in The Gambia, the intensive growing of vegetables, mainly by women for the urban market and tourist hotels, has increased in recent years.

In places such as northern Nigeria, where there is a prolonged dry season from October to June, some people, mainly men, migrate to the towns for longer periods and take up employment there. Movement of people from urban to rural areas also occurs, as in the cases of town-based traders taking manufactured items to rural villages and towns-people visiting relatives in the countryside. Such rural–urban links are very strong in many parts of tropical Africa and will be considered in more detail later.

In spite of poor roads and poor public transport systems, it is often surprising that circulation within Africa's rural areas is so widespread. Farmers, for example, will often walk long distances to get to their farms, their journey to and from work sometimes taking two or more hours in each direction. It is not uncommon for farmers to leave their village before dawn and return after dark. In areas where rotational bush fallowing is practised, the farm may be in a different location each year. Movement between rural villages is also common, particularly where markets attract sellers and buyers from surrounding areas. Rural people may also leave their village to assist with agricultural work such as cash crop harvesting in other rural areas. There are examples within tropical Africa of enforced migration in rural areas, such as the villagization programmes in Tanzania (see Case study H) and more recently in Ethiopia. Rural non-farm employment may also draw people away from their home villages. In the Eastern Province of Sierra Leone, for example, large numbers of male farmers engage in alluvial diamond mining during the dry season, returning to their villages at the onset of the rains to prepare their farms for cultivation. A lucky diamond discovery can make a valuable addition to the income of a rural family.

Population movements, whether voluntary or enforced, short- or long-distance, permanent or temporary, have had a marked effect on the distribution of population in tropical Africa.

Racial and ethnic diversity

The people of tropical Africa may be divided broadly on the basis of race into Negroid and non-Negroid, the former being much more common in sub-Saharan Africa. However, in the eastern horn semitic and hamitic groups are to be found, such as the hamitic Tuareg and Fulani in the Sahara desert and Sahel region and other groups in parts of Ethiopia and Somalia. Most Ethiopians are hamitic, but there is a strong semitic influence in culture and language. South of the Sahara and the Horn of Africa, the Negro race dominates, the characteristic features being broad noses, thick lips and black curly hair. However, amongst negroes there is considerable variation in height and facial characteristics. The Pygmies living in the forests of Zaire and Congo, for example, average only 1.35 metres in height.

Different ethnic origins are reflected particularly in language. In the continent as a whole there are some 1000 languages, divided into about 100 language groups. There is considerable language diversity in some West African countries, such as Nigeria, where no fewer than 395 different languages have been identified, and in some cases a single language, such as Yoruba, may have many dialects. Even a country such as Sierra Leone, which is smaller than Scotland, has 16 indigenous languages. In contrast, in southern and central Africa there is less language diversity, with languages such as Swahili being spoken over a wide area. In addition to Swahili only three other languages, Amharic,

Arabic and Hausa, have more than local significance. As a result the languages of the colonial rulers, notably English, French and to a lesser extent Portuguese, have become widespread and are in common use for official government business. Many governments today are keen to preserve traditional languages alongside European ones.

Religion is another aspect of the cultural make-up of Africa's people. Traditional religions with ancestral cults, magic and secret societies still exist in parts of tropical Africa, but have been eradicated or often overlain by Islam and Christianity. Islam, which entered Egypt in AD640, spread rapidly westward, but took some time to penetrate south of the Sahara. From the late-eighteenth century its influence spread into the Sahel and Savannah zone of West Africa, whilst on the eastern coast of Africa Islam flourished due to regular contact with Arab traders. Since the Second World War it seems that the spread of Islam has continued, particularly into southern parts of West Africa, which during the early colonial period were strongly influenced by Christian missionaries. In spite of possibly losing some ground to Islam, Christianity in various forms continues to be widespread in Tanzania, Uganda, Zaire and southern Nigeria, notably in the densely settled homelands of the Yoruba and Ibo. It was estimated in 1970 that there were some 65 million Christians in Africa as a whole.

The term 'tribe' is immediately associated with Africa, but many states have tried to play down tribal divisions since independence in an effort to promote national unity. In many tropical African countries, however, ethnic diversity and, in particular, tribal rivalries have been the cause of internal instability and civil war. For example, the Nigerian civil war of 1967–70 was caused by the Ibo people in the south eastern part of the country setting up a separate state of Biafra and wanting to retain valuable revenues from oil produced in the region. In Liberia, the military coup which put Master Sergeant Samuel Doe into power in April 1980 overthrew William Tolbert and the True Whig Party which was dominated by the minority Americo-Liberians who controlled much of Liberian economy and society. Doe's own overthrow and assassination ten years later in 1990 was itself the product of intense tribal rivalries as well as the leader's unpopular and harsh dictatorship. In Sudan, conflict between north and south has been due mainly to the Khartoum-based government imposing the Arabic language and Islamic Sharia law on the largely non-Muslim population of the south. In neighbouring Ethiopia, the former Italian colony of Eritrea has been fighting for independence since 1962, whilst in the province of Tigray the Tigray People's Liberation Front has also been struggling for greater autonomy for many years. Such instability and civil war has had devastating effects on fragile African economies and has caused poverty, misery and homelessness amongst a great many innocent people.

Quality of life: health and welfare

Quality of life can be evaluated by reference to a wide range of statistics, though not all are available or reliable for many tropical African countries. Per capita income, literacy levels, life expectancy, infant mortality and daily calorie intake are some of the most useful variables in evaluating quality of life. Very few countries have reliable data on incomes, unemployment registers or minimum wage levels. Since most of Africa's people live in rural areas where they are primarily engaged in food production, the bulk of which does not enter the market, it is difficult to ascertain incomes without detailed household surveys which are both difficult and expensive to undertake.

If per capita Gross National Product (GNP) is used as a measure of national wealth, then tropical African countries are amongst the poorest in the world. In fact, four out of five of the world's very poorest countries are in Africa, and Mozambique, with a per capita GNP in 1989 of $US 80, was ranked by the World Bank as the world's poorest country, followed closely by Ethiopia ($120), Tanzania ($130) and Somalia ($170) (see Table 2.1, pp. 26–7). In mainland tropical Africa the country with the highest per capita GNP in 1989 was Gabon at $2960, which is relatively small when compared with $20,910 for the USA and $14,610 for the UK. What should be remembered is that these figures are averages and therefore conceal considerable variations amongst the population. In many African countries it is common for a small proportion of the population to own a large proportion of the wealth. Small elite groups, often based in the capital cities and with good access to political decision-making, are common in many African countries. Sometimes these elites are concentrated among particular ethnic groups such as the Creoles in Freetown, Sierra Leone and the Americo-Liberians in Monrovia, both of whom once played a major role in economic and political affairs in their respective countries.

Standards of health and education are poor in most tropical African countries. Adult illiteracy is widespread and includes over 80 per cent of the population in Somalia (88 per cent), 87 per cent in Burkina Faso, 86 per cent in Niger and 83 per cent in Mali. Female illiteracy in these and

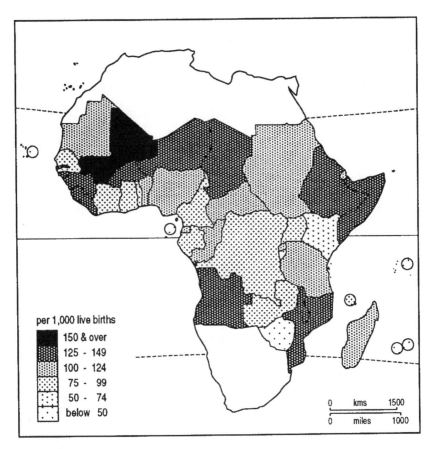

Figure 2.3 Infant mortality, 1989

other countries is even higher. Zambia and Zimbabwe are unusual in having illiteracy rates of 24 per cent and 26 per cent respectively, and their female illiteracy rate of 33 per cent in both cases is well below the average figure for tropical Africa. Health care, particularly in the rural areas, is at best rudimentary and at worst non-existent. This is reflected in low life expectancy and high infant mortality figures. Infant mortality is a particularly useful indicator, since it reveals the state of child care, the well-being of the mother and more general household conditions. A figure of over 100 deaths per 1000 live births is common in tropical Africa. Mali with 167 and Sierra Leone with 149 are the worst in Africa and are a stark contrast to figures of 10 or less for western Europe,

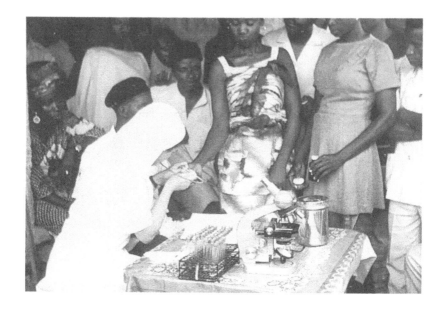

Plate 2.3 Catholic health worker taking blood samples to test for parasitic diseases in Sierra Leone.

North America and Scandinavia. Zimbabwe's figure of 46 is the lowest in mainland tropical Africa.

Daily calorie supply is another key statistic since it affects life expectancy, infant mortality and susceptibility to a wide range of diseases. Calorie supply in the countries of western Europe and North America is well over 3000 calories per day. For example, the USA has a figure of 3666 and the UK a figure of 3252. In many tropical African states the figure is below 2000, for example Mozambique with 1632 calories per day and Ethiopia with 1658. Some countries have actually experienced a fall in calorie supply between 1965 and 1988, most notably in Chad (from 2374 to 1852) Mozambique (1704 to 1632) and Ethiopia (1802 to 1658), all of which are countries which have suffered from civil war as well as drought.

The United Nations Development Programme (UNDP) in its Human Development Report 1990, suggests that the key indicators in measuring human development are life expectancy, literacy and command over resources needed for a decent living. The latter, which is the most difficult to measure, requires data on access to land, credit, income and other resources. However, since data on many of these variables are scarce, per capita income is possibly the best indicator. Using life expectancy, literacy and income, the UN has produced a Human Development Index (HDI) with countries categorized into low, medium and high human development. The country with the lowest HDI in 1990 was Niger (0.116), and nine out of the ten countries in the world with lowest HDI's were in tropical Africa. In fact, when the table of countries with under one million people is examined, there are three small African countries with HDI's even lower than Niger, namely Djibouti (0.073), Guinea-Bissau (0.074) and The Gambia (0.094). No tropical African country falls into the high human development category and only two, Gabon (0.525) and Zimbabwe (0.576), fall into the medium human development group.

Amongst tropical African countries, Zimbabwe may be singled out as a country which has made considerable progress in human development in recent years (see Case study M). At Independence in 1980 the Zimbabwe government made a conscious effort to reduce inherited racial inequalities and improve the position of the poor. The share of defence in total government expenditure was reduced from 44 per cent in 1980 to 28 per cent in 1984, whilst the share of education and health rose from 22 per cent to 27 per cent. Within the education budget, the share of primary education rose from 38 per cent to 58 per cent over the same period, involving a doubling of real per capita expenditure on primary education. Health care became free for the vast majority of the population and an increasing proportion of the health budget was devoted to preventive health care. The child vaccination programme was expanded and a Department for National Nutrition established for nutrition and health education, nutrition surveillance and supplementary feeding programmes for children. As a result, infant mortality and malnutrition declined, whilst primary school enrolment rose rapidly.

Zimbabwe's achievements in improving the quality of life amongst its people are impressive, but Zimbabwe is very much the exception rather than the rule in tropical Africa. The majority of people in most tropical African countries are impoverished, unhealthy, undernourished and uneducated. There is much scope for improvement, but in some countries this is hindered by rapid population growth and inappropriate development policies. The link between population and development is of great importance in all countries, and in certain tropical African states very urgent attention is needed.

Population and development

Rapid population growth is often seen as the major barrier to Third World development and the improvement of living standards. There are two important aspects to be considered however;

1 First, it is unwise to generalize, since in tropical Africa there is much variation from one country or region to another in terms of population pressure on available resources.
2 Second, the reasons behind rapid population growth must be understood. With such high infant mortality in many African countries there is an incentive for parents to have more children in the hope that some will survive. Another factor is the early age of marriage and first child-bearing, though this should be considered alongside the relatively short life expectancy. In some societies polygamy is common and status is attached to having a number of wives and their offspring. Children provide important help in farming and domestic duties, such as fetching water and firewood. Finally, in countries where there is no social security system, parents depend on their children to look after them in old age. In most African societies old people are respected members of the community and Africans fail to appreciate why in the so-called 'developed' world we have special homes for old people quite separate from their families.

These facts must be borne in mind when discussing population growth and control. Some would say African people are poor because they have too many children. Others argue that as living standards and education improve family size will decline, as it has done in the countries of western Europe and North America since the late-nineteenth century. Although Africa by world standards is sparsely populated, with only 18.2 persons per square kilometre in 1985, compared with 101 per sq. km in Europe and 102 per sq. km in Asia, over half of Africa is uninhabitable and much of the land currently under cultivation is marginal. However, it is estimated that by 2025, on the basis of current growth rates, Africa's population density in habitable areas would be the same as that of present-day Europe. There is, therefore, an urgent need for all tropical African countries to consider taking steps to reduce population growth, yet many still have only weak or non-existent family planning programmes. However, in the late 1980s and early 1990s a number of African countries drafted national population policy statements, notably Liberia (1988), Nigeria (1988), Senegal (1988), Sudan

(1990) and Zambia (1989). Many other countries are in different stages of formulating a population policy. The tension between population and resources in the 1980s, together with a better understanding of demographic mechanisms, have helped family planning to be no longer perceived as a Western value imposed upon African societies, but, on the contrary, as a way to reinstate traditional African birth-spacing patterns.

Case study B

Overpopulated Kenya

Kenya, with one of the most rapid population growth rates in tropical Africa, and indeed the world, was one of the first countries in Africa to adopt a national family planning programme in 1967. Kenya's average annual population growth rate between 1980 and 1989 was 3.8 per cent. In 1984 the President held a Conference on Population at which he reaffirmed and intensified the country's commitment to reducing population growth. The seriousness of the situation was again recognized by the Kenyan government in 1986, when it referred to population growth 'overwhelming the economy's capacity to produce and provide for its people'. A population of 20.6 million in 1986 was projected to rise to 35 million by 2000 and 83 million by 2025. This rapid growth is partly due to infant mortality being halved between 1948 and 1985 and life expectancy increasing from 39 in 1955 to 59 for a Kenyan born in 1989. At 7.7 per cent per annum, the rate of urban population growth has been even higher than the general population growth rate. Large families are often associated with the low status of Kenyan women for whom one of the few ways of improving status is by producing a large number of children, a phenomenon which is not restricted to Kenya. In recent years there has been an increase in surgical contraception, contraceptive distribution by non-professional workers and the setting up of purpose-built family planning clinics in urban areas.

Kangundo village in Machakos District had a population of 54,147 people with 10,163 households at the 1979 census. The density was 406 people per sq. km and the agricultural land only 0.2 hectares per person. The population is still increasing, with

Case study B (*continued*)

people concentrating on more productive areas of land, living on hillsides and suffering from chronic seasonal food shortages. Holdings are getting smaller, cultivation is more intensive and there is much unemployment. Many people are now landless and, with no space to rear cattle, the lack of manure has resulted in the land losing its fertility. Expansion and movement to other rural areas is restricted by different ethnic territories, so rural to urban rather than rural to rural migration is common.

The Greenbelt Movement has encouraged women, coming together under the aegis of the Family Planning Association of Kenya, to plant trees for economic crops and fuelwood. Most women in the community are aware of the need to limit the size of their families, but the average number of children per woman in the late 1980s was six, falling short of the Kenyan government's target of four. Each community-based distribution agent is responsible for distributing contraceptives and information on family planning to around 1000 households. The family planning network is having some success; the 1989 Kenya Demographic and Health Survey found that 40 per cent of married women in rural Machakos District were using some form of contraceptive method. The survey also found that in Kenya as a whole between 1984 and 1989 the total fertility rate fell from 7.7 to 6.7, a reduction of roughly one child per woman. Women are also marrying slightly later, though 50 per cent of women still marry and become mothers before they reach the age of 20. Perceptions of the ideal family size also changed from 5.8 children to 4.4 children between 1984 and 1989. The authors believe a major factor behind the change in views is the increasing cost of education which has changed the balance of benefits and costs of raising children. Other factors, such as the Government's commitment to reducing population growth, the increase in family planning services, AIDS, the 1984 drought and a reduction in infant and child mortality, have also played a part. The study also proposes reducing the social costs of family planning, a greater sensitivity to local needs and an increased role for non-governmental organizations in the delivery of family planning services.

Case study B (*continued*)

Kenya is making steady, but slow, progress in reducing population growth. A more integrated approach is needed, where family planning workers and agricultural extension officers work together to raise food production, living standards and the status of women, since these and many other factors are all interrelated.

Case study C

Underpopulated Mali – feeding the people

Mali is a vast landlocked state in the semi-arid savannah-sahel zone of West Africa, with an area of 1,240,192 sq. km, larger than the combined areas of the UK, France and Germany. Whereas the latter three countries have a total population approaching 200 million, Mali's population in 1989 was estimated at only 8.2 million, with an average population density of 6.6 people per sq. km, one of the lowest in tropical Africa. The vast majority of Mali's people live in the wetter southern half of the country, where the main towns are located, notably Bamako, the capital (646,000 people), Segou (70,000), Mopti (60,000), Kayes (60,000), Sikasso (50,000) and San (25,000). Mali's most famous town, Timbuktu, was once an important trading, religious and educational centre in the pre-colonial period, with an estimated population in the fifteenth and sixteenth centuries of 25,000. However, its fortunes have since declined and in the 1960s its population fell to around 7000.

Mali is a very poor country with over 80 per cent of the population living in rural areas and 85 per cent of the labour force employed in agriculture. In 1989 Mali had a per capita GNP of only $270, life expectancy was 48 years and the country had the highest infant mortality rate in the world, 167 deaths per 1000 live births. Daily calorie intake was only 2181 per capita in 1988, well below the figure of 3149 for the UK. If statistics such as life expectancy, literacy and GDP per capita are aggregated, Mali had the world's sixth lowest human development index in 1989.

Undoubtedly the key to the life and the future development of

Case study C (*continued*)

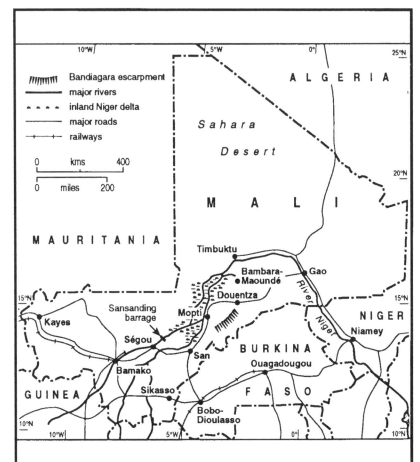

Figure 2.4 Mali

Mali is water. The country's greatest asset is the river Niger, rising in the Fouta Djallon plateau of Guinea to the south, flowing northwards through Bamako towards Timbuktu, then bending south through Gao to the border with Niger, a total distance within Mali of 1700 km. The river provides a year-round navigable routeway through the country, which has been used for centuries for transport and trade. However, in 1985, at the end of a long

Case study C (*continued*)

period of drought, the Niger almost dried up near Gao. North of the Niger bend, population and vegetation become progressively more sparse. Cultivation is limited and pastoralism is the dominant food production system. Further south, and downstream of Bamako, the river loses speed and braids into a series of channels across a vast area known as the Inland Niger Delta, one of the natural wonders of West Africa. With an area of 103,000 sq. km, the delta is not far short of the combined areas of Belgium, The Netherlands and Switzerland. During the rainy season from June to October, the plains of the delta are flooded. Ancient settlements such as Mopti and Djenne stand on islands linked by causeways above the flood plain. Land watered by the Niger flood has long been used by Bambara farmers for the cultivation of crops such as rice, millet, sorghum and maize, whilst the river channels are important fishing grounds for the Bozo and Somono fishermen. Fulani and Tuareg pastoralists also make use of the delta for grazing their herds.

With such a small and poor population, living in a vast country endowed with the resources of the Inland Niger Delta, one is tempted to ask the question, 'could the land be used more profitably to raise food production, improve domestic living standards and possibly earn valuable foreign exchange through agricultural exports?' This question was considered by the French colonialists who ruled Mali (then called French Soudan) until 1960. They saw great potential in the country, and believed the delta could become 'the rice granary of West Africa'. In the early years of the French colonial period, cash crop production in Mali concentrated on cotton for export rather than rice for domestic consumption, with a small cotton scheme near Segou at the southern end of the delta. After the First World War in 1919–20, the French proposed that one million hectares of irrigated land, more than half under cotton and some under irrigated rice, should be developed on the left (west) bank of the Niger. As the area was underpopulated, it was estimated that at least 300,000 immigrants would be needed to work on the scheme over a period of 25 years. In 1932 the Office du Niger (ODN) was created to coordinate operations, and it

Case study C (*continued*)

planned to build two barrages, including a major one at Sansanding, and a series of related canals and irrigated areas. ODN was also to install rice mills and cotton ginneries, build settlers' villages and provide medical and education services. Work started in 1934 on the Sansanding barrage, together with a navigable by-pass and associated canals. Although the canals were completed a year later, the barrage was not finished until 1947. In the meantime in 1935, 6600 ha were planted with rice and 8000 ha with cotton.

During the Second World War the French Vichy government allocated money to irrigate a further 200,000 ha mainly for cotton production. The ambitious plans even envisaged the building of a railway across the Sahara desert to provide a direct route to the Mediterranean. However, by 1945 ODN had little to show for the massive expenditure. Despite the pilot schemes, inadequate knowledge about soil and agronomic conditions had resulted in a general failure of the initial stages of the irrigation project before the Sansanding barrage was even completed. The colonists who had arrived had difficulty tilling the land with the tools provided, and many of the settlers were Mossi people, who came from a very different environment where floodland cultivation was not particularly significant. They often brought large numbers of relatives and hired labourers to work on their farms, with the result that overcrowding occurred in some areas. Although much emphasis was placed on cotton production, rice actually proved to be more profitable, with good prices and markets available in both Mali and neighbouring Senegal. In 1945 it was decided to stop all further expansion of the scheme and undertake more detailed research. By 1957 the popularity of rice had increased further and the crop occupied 60 per cent of the land, whilst cotton had reduced to only 15 per cent. Despite poor performance on the scheme, investment in ODN continued to flow. During the period from 1945 to 1957, ODN received 30 per cent of the colonial agricultural development funds, whereas the rest of the Agriculture Service received less than 4 per cent of the annual colonial budgets.

Case study C (*continued*)

Following independence, ODN was transferred to the Malian government in 1961. By 1962, thirty years after its creation, the scheme's total area was still only 50,000 ha. Total expenditure by the ODN was estimated at 1962 values to be £36.5 million, or £730 per hectare. During the 1960s, rice production was affected by marketing problems, infestation by wild rice and attacks by quelea birds. Fertilizer costs escalated since, with the exception of phosphates, they had to be imported. In spite of all these difficulties, in 1986 it was felt that ODN was worthy of yet more financial assistance, with emphasis now being placed on the production of rice and other food crops with the aim of achieving national self-sufficiency. It was decided that ODN should cease cotton cultivation altogether. A major rehabilitation programme of ODN was therefore initiated which, at a cost of about £23 million, aimed to rationalize the organization's management and increase the total cultivable area to more than 100,000 ha, of which 46,730 ha were to be planted with rice and 7700 ha with sugar cane by 1989. The programme was supported by loans from the International Development Association, France, Germany, the Netherlands and the EC. Output of raw sugar increased from 10,000 tons in 1983–4 to 18,100 tons in 1986–7, and to an estimated 20,000 tons in 1987–8, thus almost meeting domestic demand. However, yields of rice still remained low, largely due to poor rainfall. In 1988 a further programme was inaugurated, with the aim of improving the irrigation network for rice cultivation. The cost of this programme was estimated at £50 million and is being supported by the World Bank and European donors. After the drought years of the early 1980s, Mali's production of rice increased in the second half of the decade from 187,200 tons in 1985–6 to 294,000 tons in 1990–1, at last meeting domestic requirements. Seed cotton production also increased from 175,100 tons in 1985–6 to 255,000 tons in 1990–1, largely due to development programmes in the southern regions, which largely escaped the 1982–4 drought.

In spite of massive investment in ODN over six decades, the performance of irrigated agriculture in Mali has been disappointing. In the early days too much emphasis was placed on cotton

Case study C (*continued*)

production to supply French industries and not enough attention given to feeding the Malian population and exporting food surpluses to other food deficit countries in the region. Post-independence governments have favoured institutional solutions to agricultural and rural development supported by foreign aid, rather than policies designed to promote more self-reliant rural development. Poor management and maintenance of the ODN scheme, combined with inadequate incentives to tenant farmers are major factors to blame for its uninspiring record. The Inland Niger Delta still has much untapped potential. Perhaps the focus and style of future development strategies should change to encourage community-based smallholder schemes rather than large complex institutions such as ODN.

Key ideas

1 An absence of recent and reliable population and other data in many tropical African countries makes the analysis of existing patterns and processes difficult and hinders future development planning.
2 In general terms tropical Africa is sparsely settled, but there are some areas of dense rural settlement as well as rapidly growing urban areas.
3 Population growth rates in tropical Africa are amongst the highest in the world.
4 The movement of people over varying distances and for varying periods of time is a long-established feature in many parts of tropical Africa. This process was accelerated with the effects of colonial policy, notably due to the creation of large urban centres, mining enclaves and areas of cash crop production.
5 The countries of tropical Africa have some of the world's lowest levels of human welfare.

3
African environments

Interest in environmental issues

The environment, or rather the environments, of tropical Africa have been much in the news in recent years and have been blamed for many of Africa's problems. Diverse, inhospitable and unpredictable are descriptions of African environment which are used as much today as they were over a century ago when the European powers were examining the possibilities of colonizing the 'dark continent' (see Figure 3.1). This criticism of environment is unwarranted, since what may be perceived as a purely environmental issue in fact often has complex social, economic and political elements to it. For example, famine can rarely be explained in purely environmental terms. Civil war, economic instability and government policy have played an important role in contributing towards famine in many cases. The conditions in Somalia in August 1992 testify to this. Widespread starvation was largely brought about by civil war that destroyed the capital Mogadishu and destabilized social life and food production in even the remotest communities.

However, given the low level of technology and development in much of tropical Africa, people are more vulnerable to the vagaries of environment than in richer, more technologically advanced countries. In a region where well over half the population works on the land, a year with below average rainfall or a plague of locusts may have a profound effect on food production and social welfare.

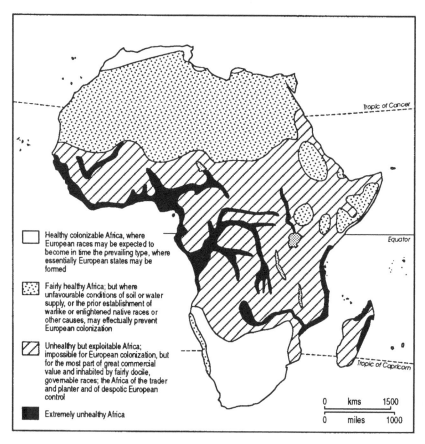

Figure 3.1 Colonizability of Africa
Source: Based on Sir H. H. Johnston, *A History of the Colonisation of Africa by Alien Races* (CUP, 1905, facing p. 275).

The diversity of environment

Within tropical Africa there is a great diversity of environments ranging from the great expanse of the Sahara desert in the north, through the semi-arid areas of the Sahel, to the savannah grasslands, rainforest and mangrove swamps. In certain places environment is affected by altitude, such as in the Ethiopian and Kenyan highlands, where it is cooler and a range of plant types more typical of temperate areas is to be found. Even within a single country, such as Sudan, the largest state in Africa, there is an impressive range of environments from the moist forest and

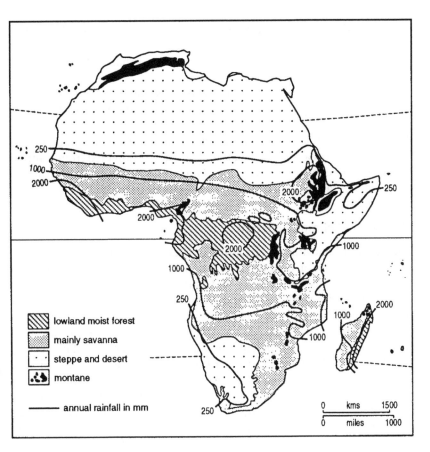

Figure 3.2 Vegetation and rainfall

vast swamps of the Sudd in the south, to semi-arid grasslands, and ultimately to desert in the far north.

Climate, and particularly rainfall, is the dominant factor affecting environment. Indeed it might be said that water is the key to much of Africa's survival and future development. Despite the great diversity, there are three major environmental types in tropical Africa determined largely by rainfall quantity and variability (see Figure 3.2):

1 Tropical rainforest – tall dense forest found in areas with more than 1400 mm of rainfall and with no drought period;

2 Savannah grassland – with varying densities of tree cover, found in

areas with three to eight months drought and heavy rainfall at other times;
3 Steppe and desert – where rainfall averages less than 400 mm per year.

Other environments are variations of these three basic types or are intermediate between two of the three. For example, the Sahel (or Sahelian) environment, about which much has been written in recent years, is an intermediate environment between savannah and desert. In fact in West Africa, the savannah zone is often subdivided from south to north with decreasing rainfall, into Guinea savannah, Sudan savannah and Sahel savannah, each having a distinctive range of plant species.

Rainfall and climatic change

The quantity and timing of rainfall is vital throughout tropical Africa. In general terms a belt of rainfall shifts southwards from equatorial latitudes towards the Tropic of Capricorn between November and April, and northwards towards the Tropic of Cancer between May and October (see Figure 3.3). In January, the highest rainfall totals are to be found in Madagascar and southern Zaire, whereas in July it is regions to the north of the equator which receive most rainfall, such as the Ethiopian highlands in the east and the coastal states of Liberia, Sierra Leone and Guinea in the west. But as far as human activity is concerned, it is the effectiveness of the rainfall which matters. Whether rainfall is adequate for crop growth depends on the rate at which moisture is taken up by evaporation from the surface and by transpiration from plants.

There is also a considerable degree of variability in Africa's rainfall, particularly in desert, semi-desert and savannah areas. There are those who believe that the climate of tropical Africa is changing and becoming drier, leading for example to the drought periods of the last three decades. Between 1968 and 1989 in the Sahelian countries of West Africa, every annual rainfall total was below 'normal', if indeed 'normal' can be defined. The early twentieth century was also a relatively dry period, with only 1906 and 1909 being above normal in a 21-year period. In Sudan, where relatively good data exist, two major periods of dryness between 1900–19 and 1965–85 were separated by a generally wet period from 1920–64. Such cycles of wet and dry periods have been detected in many parts of Africa, but whether there is a long-term drying-up of the climate associated with global climatic change is less

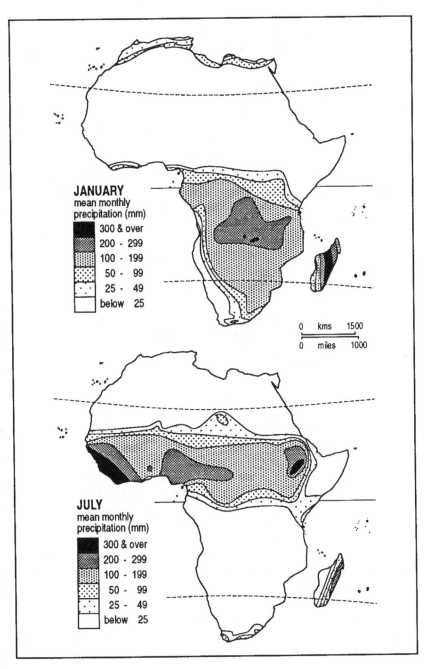

Figure 3.3 Precipitation in January and July

easy to prove. Much has been written in recent years about the process of 'desertification', associated with periods of drought.

Is desertification a myth?

The common image conjured up by the term 'desertification' is one of an advancing Sahara desert moving south, smothering villages and destroying farmland and pasture once and for all. It is often suggested that this process is the result of a long-term decline in rainfall, exacerbated by unwise human practices such as overgrazing, burning and deforestation.

Reports of an advancing Sahara desert are by no means new. In fact, at a meeting of the Royal Geographical Society on 4th March 1935, Professor E P Stebbing, Professor of Forestry at the University of Edinburgh, delivered a much-quoted paper entitled 'The encroaching Sahara: the threat to the West African colonies'. Stebbing, reporting on a recent visit he had made to northern Nigeria, spoke of the advancing desert and proposed, as one might perhaps expect from a Professor of Forestry, that two massive forest belts be planted, one through what is now Burkina Faso and northern Nigeria to Lake Chad, and another further south from Ivory Coast, through Ghana (formerly Gold Coast) to Jebba or Minna in Nigeria's middle belt. It was estimated that up to 16,000 million trees would have to be planted, though Stebbing favoured the use of existing woodland which would be closed and protected from farming, fire and grazing.

The events which perhaps did most to popularize the use of the term 'desertification' in recent years were the two droughts in Africa's Savannah-Sahel zone: 1968–74 and 1979–84. Following the first drought and the associated media coverage, the United Nations Environment Programme organized a World Conference on Desertification (UNCOD) in Nairobi in 1977 to consider the extent and character of the problem and to propose measures to combat it on an international scale (see Figure 3.4). UNCOD defined desertification as 'the diminution or destruction of the biological potential of the land . . . leading ultimately to desert-like conditions. . . . Desertification is a self-accelerating process, feeding on itself and as it advances, rehabilitation costs rise exponentially' (UNCOD, 1977).

But what is the reality of the situation? Evidence from field observations suggests that desertification has been grossly over-emphasized, largely because of insufficient investigation on the ground as to how environments, societies and food production systems respond to periods

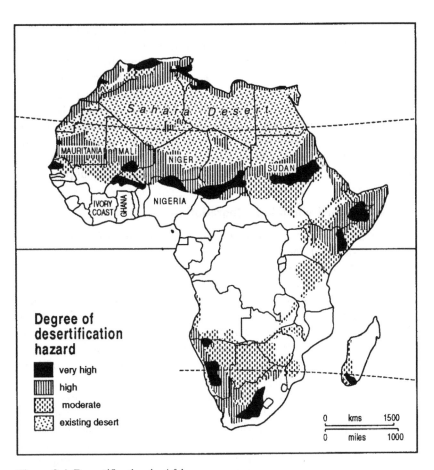

Figure 3.4 Desertification in Africa

of drought. A UN Environment Department Working Paper written by Ridley Nelson in 1988 said, 'Contrary to popular belief, the extent of desertification is not at all well known . . . there is extremely little scientific evidence based on field research or remote sensing for the many statements on the global extent of the problem.'

A major concern is that of defining 'desertification'. The term has been used widely to raise public awareness, but should now be used more cautiously to describe land which has once and for all been removed from productive use. This should be differentiated from areas which have suffered 'land degradation', which, though severe in some cases, with care and two or three good rainy seasons might return to their former status. It is not that desertification does not exist, but rather its existence over space and time is much less than the popular image. The fringe of the Sahara desert, for example, might be likened to an ebbing and flowing tide, moving southwards during dry periods and northwards when rainfall is plentiful. Pockets of degraded land may occur at any time for a variety of human and environmental reasons, but the process of degradation is likely to accelerate during dry spells. However, claims that the Sahara is expanding at some horrendous rate are still made despite the absence of detailed evidence to support them.

Carrying capacity and land degradation

Land degradation of one sort or another is a feature of many environments throughout the world. But in some marginal environments of tropical Africa features such as soil erosion, gullying and deforestation have received much attention. Frequently, local farmers and pastoralists are blamed for mismanaging their land and taking too much out of it. Overgrazing, the burning of vegetation to stimulate new growth and the use of timber as fuelwood are among the practices identified as leading to land degradation. It is often suggested that these practices indicate the presence of too many people and animals on an area of land, exceeding the land's carrying capacity.

William Allan, writing in 1949, suggested that for every area of land on which people are growing crops or rearing livestock, there is a human and animal population limit which cannot be exceeded without setting in motion the process of land degradation. Allan called this limit the 'critical population' or 'carrying capacity' for that system of land usage. He suggested that the possibility of expressing carrying capacity numerically allows a much more precise approach to the problems of African agriculture. Allan argued that the calculation of carrying capacities for traditional African systems can be achieved, but requires a close study and clear knowledge both of the land and the systems employed on it. To obtain the 'magic' figure he fed the following data into his calculation; data on vegetation and soil type, the total area of each of these types, the cultivation and fallow periods for each of these types based on traditional land use practice, the cultivable percentages allocated to each type for the system of production and the total area of land required per head of population. He was then able to arrive at a figure for the carrying capacity of specific areas (see Figure 3.5).

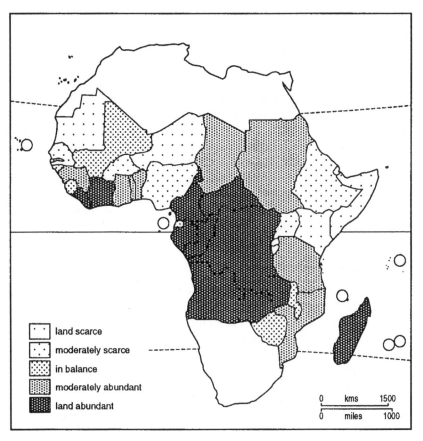

Figure 3.5 Potential population support capacity

But how easy would it be to calculate such a figure for all or most of
tropical Africa, and what use would it be? Allan recognized that estimates are strictly relative to specific areas and to the systems employed
on them and that the concept of carrying capacity assumes a more or less
ideal distribution of population in relation to soils. He admitted that the
two main weaknesses of methods used in arriving at carrying capacity
are the problem of obtaining accurate figures for the percentage of
cultivable land and the amount of labour needed in surveying and
sampling farms. In reality, the sort of data required for such an exercise
are just not available in many African countries, and detailed field surveys would be both prohibitively expensive and practically impossible.

Even if such a carrying capacity figure were obtainable, it might only be useful for one particular area at a certain point in time. As we have seen, environmental factors, such as climate, change frequently and traditional food production systems are also by no means static. Indeed, with a system of nomadic or semi-nomadic pastoralism, the movement of people and animals in search of pasture and water would surely make a figure of carrying capacity worthless. Even with more sophisticated technology than Allan had at his disposal, such as air photography and satellite imagery, it is difficult to see the value of calculating carrying capacities with such a dynamic set of variables and a lack of detailed ground-level surveys. Although the concept of carrying capacity may be of limited practical use for African conditions, the underlying principles on which it is based are, however, perhaps worthy of consideration in any resource management programme.

Adapting to environment

There is a need to appreciate people–environment relationships if the threat of famine and land degradation is to be reduced in future development programmes. There is now, for example, plenty of accumulated evidence to show how pastoral and farming communities in marginal semi-arid and arid regions of tropical Africa have devised, over many generations, mechanisms to cope with harsh environmental episodes such as drought, which is endemic in these regions and is a key element in indigenous folklore and oral histories. Food production systems frequently possess a high degree of structural resilience and adaptation to environment.

A survey undertaken in 1982–9 of households who migrated from famine-affected communities in Northern Darfur to the provincial capital, El Fasher, in western Sudan, revealed that some groups were able to stay longer in their villages before migrating because they adopted a range of survival strategies. The most popular strategies reported were the use of alternative income-generating activities such as the sale of wild foods, transport hire and wage labour. Other survival strategies included changing cropping patterns, cultivating larger areas, multiple cropping, the use of different pastures and the sharing of resources. A detailed long-term study of communities in the far north of Nigeria also shows a wide range of adaptive mechanisms for coping with recurrent droughts. Productive systems remained resilient, the communities survived and during the 1980s even managed to export grain to the market.

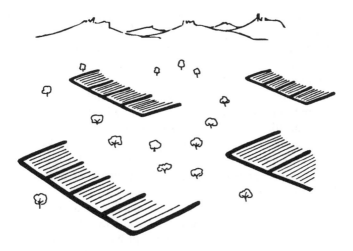

Figure 3.6 Traditional teras cultivation in Sudan

There are many examples of indigenous methods of soil and water conservation in tropical Africa. For example, in the Ader Doutchi Maggia region of Niger where rainfall is between 250 and 450mm, lines of stones have been laid out by people living on the barren plateaux to conserve water and trap windblown sand. After five or six years sufficient sand has usually been deposited to start cultivation and the addition of millet stalks and manure help to raise fertility in the new plots. Stone lines are found elsewhere in the Sahel region of West Africa. In Yatenga, north-western Burkina Faso, rock barriers, bundles of stalks and branches placed across the direction of water flow help to prevent run-off and soil erosion. Small stone dykes or 'diguettes', about 30–40 cm high and built along the contours, are even more effective in stabilizing the soil (see Case study D).

Teras cultivation is widely used on the clay plains of eastern Sudan. The *teras* is an earthen bund 35–40 cm high with a base of 1.5–2.0 m, surrounding three sides of a cutivated plot of up to 3 ha (see Figure 3.6). The bunds impound run-off from the plains and the system produces a chequer-board landscape of cultivated rectangular plots interspersed with areas of uncultivated plain. Sorghum is the main crop, with water melons planted on the bottom bund of the open rectangle.

The Dogon people in the area south of Mopti in Mali use similar systems of earthen ridges and basins, but unlike the *teras* of Sudan, the

Dogon do not leave an open end to the rectangle. In fact the Dogon have developed an impressive range of soil and water conservation techniques including stone lines, stone bunds and earth mounds. The latter, as well as slowing down run-off, also act as small compost heaps and help maintain soil fertility.

In Sierra Leone, which has a much higher rainfall than the Sahel, traditional methods of stick and stone bunding in the mountains of the Freetown peninsula are very effective in preventing soil erosion and require little capital or extra labour input. Bundles of sticks are placed across the slope in continuous lines or bunds and pegged into the ground. The bund slows down run-off and valuable sediment is deposited behind the bunds.

On steeper slopes in many parts of tropical Africa are to be found intricate terrace systems which were often constructed centuries ago and are carefully maintained to this day. They provide yet another example of soil and water conservation strategies. For example, the Mafa people in the Mandara Mountains of northern Cameroon, where rainfall is 700–1100 mm, have developed a complex system of terracing on the steep slopes. Fertility is maintained with the addition of household waste, animal manure and crop residues, whilst trees growing on the terraces help to stabilize the soil.

Managing forest resources

There has been much concern in recent years about the loss of trees during drought periods due to the need for firewood in rural and urban communities. In Kenya as a whole, fuelwood accounts for 53 per cent of the total energy consumption, but the figure is much higher in the rural areas where alternative supplies of fuel are not available. The situation is similar in many other African countries and there is an urgent need to encourage the preservation and enhancement of valuable forest resources. The reduction of woody plants can also result in soil erosion and deterioration as the surface is exposed to rainsplash and hard crusts can soon form on the bare ground.

It is often assumed that traditional peoples have mismanaged their forest resources, and deforestation is seen as one of the main causes of land degradation and desertification. However, as with soil conservation, there is evidence that many communities manage their trees carefully and have long been aware of the need to do so. In the savannah region of northern Nigeria, a study in the 1980s of woodfuel production around the large city of Kano revealed that the area adjacent to the city itself, with the highest rural population densities (up to 500 persons per sq. km), and closest to the urban woodfuel market, had not only the highest tree densities, of 12.3 trees per hectare, but also recorded an annual increase in tree numbers between 1972 and 1981 of 2.4 per cent annually. Further away from the city, there had also been a significant increase in tree density of 26 per cent between 1972 and 1985. The volume of timber per hectare is actually greater in farmland than in either shrubland or forest reserves because the trees are much larger. A single silk cotton or kapok tree, for example, may contain as much wood as several hundred trees of a smaller species with a shrubby growth habit. Farmers in the Kano area are well aware of the importance of trees, since they provide valuable shade and a wide range of economic crops and foodstuffs.

Protection of trees, as valuable elements of the environment, has been a feature of many rural communities for generations. In Niger, the sultans of Zinder at the end of the nineteenth century had a policy of executing anyone who cut down certain trees which provided important food supplies. In Mali, there is widespread preservation of tree species which have fruits which are valuable for sale or for household consumption. The protection of such species is an important example of indigenous silviculture. It shows that peasant farmers can and do take initiatives in managing their environment through protection and rehabilitation.

Pests and diseases

Once called 'the white man's grave' by the early colonialists, tropical Africa is still plagued by a wide variety of pests and diseases, a large proportion of them environment-related. In many countries infant mortality remains unacceptably high and life expectancy much too short. Inadequate and poorly structured diets result in greater susceptibility to a wide range of health problems. Amongst children, diarrhoea and measles claim many victims. Vitamin A deficiency may lead to blindness and infection (especially after measles), whilst iron deficiency often leads to anaemia, causing weakness and additional risks for new-born children. Infections associated with malnutrition are more likely to affect poor households with limited food availability, low income, restricted access to water for drinking and washing and limited health care. Pellagra, beri-beri, rickets and kwashiorkor are examples of nutrition-related diseases, to which children are particularly vulnerable.

A number of widespread diseases in tropical Africa are associated with water. Malaria, for example, is transmitted by various species of mosquitoes which breed close to stagnant or slow-moving water. The disease, which probably affects 200 million people throughout tropical Africa, weakens the victim, and is more common during the rainy season when farmers are busy producing the household's food. Mosquito larvae can be killed by draining swamps and spraying pools with insecticides such as DDT, and sprays may also be used inside dwellings. However, mosquitoes are becoming resistant to certain chemicals and to anti-malarial drugs and the incidence of malaria has actually increased in recent years. Mosquitoes also transmit dengue fever and the virus called yellow fever.

Schistosomiasis or bilharzia is a water-based disease transmitted by snails. It is spread through worm eggs expelled into urine or faeces into stagnant or slow-moving water. Larvae hatch out of the eggs and penetrate the snails. The larvae are subsequently released into the water and can penetrate the skin of people who enter the water. Ghana and Sudan have amongst the highest bilharzia infection rates in tropical Africa, and the introduction of irrigated farming seems to have encouraged its spread. The disease is rarely fatal, but it considerably lowers resistance to other infections. A number of drugs are now available for treating sufferers, but better standards of hygiene in the use of water could have the greatest effect on reducing the incidence of bilharzia.

Onchocerciasis or river blindness is another disease associated with water and is transmitted by the black fly. The disease is endemic in much of tropical Africa and is a major cause of blindness. The black fly lays its eggs on rocks and plants in turbulent water and the flies congregate near natural or artificial falls and rapids. It is near these areas that the victim is most likely to be bitten. In West Africa there is an Onchocerciasis Control Programme in eleven countries and biodegradable larvicides have been sprayed by aircraft onto breeding sites. Other major causes of ill health often associated with water are a variety of parasitic worms, notably the hookworm and guinea-worm.

The tsetse fly, a large brown fly, is the main vector of trypanosomiasis which causes sleeping sickness in humans and nagana in animals. About 190 million square kilometres within 14 degrees of the equator are infested with the tsetse fly, mainly in sparsely-settled savannah woodland. The distribution of human settlement and the presence of cattle, horses and most other domestic animals have been severely restricted in many parts of Africa through the prevalence of the tsetse fly.

AIDS, the greatest cause of concern on the world public health scene today, is rife in many parts of tropical Africa. It is twice as common among men as women. In Uganda, Tanzania and Kenya, the proportion of people who are HIV-positive is as high as anywhere in the world and rising rapidly. Although AIDS is a major problem of the towns, in Uganda it is also prevalent in the rural areas. In central Africa the countries of Congo and Zaire have been badly hit by the epidemic. In Kinshasa, Zaire in 1988 it was thought that 3–8 per cent of the population was HIV-positive. AIDS is causing a massive health care problem for already stretched and inadequate medical facilities.

Finally, one of the major pests of Africa is the locust, a species of large grasshopper which congregates in swarms. They are often a major problem after drought periods and can destroy vast amounts of growing crops. In 1985, after several years of drought, many parts of tropical Africa were affected, and by 1986 locusts were reported to be affecting Tanzania, Malawi, Zambia and eastern Zaire.

What future for the environment in tropical Africa?

What is clear from a study of tropical African environments and the events of the last three decades is the variable and unpredictable nature of environmental events and the vulnerability of many of the poor people who live off the land. Recent instances of drought and famine in tropical Africa have provoked much concern amongst the major charities and relief organizations and a considerable amount of detailed investigation has ensued. As a result, certain things have become much clearer, for example the existence of complex coping mechanisms within rural communities in marginal environments for surviving through difficult environmental events. Also, the multitude of ways in which local people have adapted to their environment and have taken steps to maintain and strengthen their environmental resource base is impressive.

What has also become clearer is that where food production and social systems have been constrained for one reason or another, and where the efficacy of adaptive and coping mechanisms has been reduced, tensions are more likely to develop, leading to events such as famine and land degradation. Political policies and development programmes have done much to destabilize traditional lifestyles. Drought need not necessarily lead to famine, but the destabilizing effects of civil war in countries such as Chad, Ethiopia, Mozambique, Somalia and Sudan have been only too evident and many lives have been lost as a result.

Colonial policies introduced in the late-nineteenth and early-twentieth centuries have also constrained people and their livelihood systems and thus increased their vulnerability. For example, the introduction of groundnut cultivation for export by the French in Senegal and by the British in northern Nigeria took little notice of traditional ways of life. The migratory routes of pastoral groups were disrupted, whilst farmers were encouraged to grow groundnuts rather than spend their time growing food crops. The introduction of taxation by the colonial powers further encouraged farmers to grow groundnuts for cash and also resulted in the large-scale migration of young men to towns and plantations in search of money to pay taxes. The migration of Mossi men from Burkina Faso to the coffee and cocoa plantations of Ivory Coast and Ghana was initiated in the colonial period and goes on to this day. Whilst the young men are absent, much of the village work must be done by women. There is no shortage of examples from tropical Africa, past and present, of government policies and development schemes which have undermined traditional ways of life, causing social and environmental problems.

Of greater value to the rural poor in marginal environments are some of the initiatives which have come from a number of non-governmental organizations (NGOs) working with the rural people at grassroots level and aimed at reducing vulnerability. The establishment of cereal banks in Burkina Faso and Mali by Oxfam since the late 1970s is an example of this. In Niger and Mali, CARE has collaborated with government forestry services in re-forestation, soil conservation and agroforestry projects.

Since the disastrous droughts and famines of the mid-1980s, there has been a boom in the business of predicting drought and famines before they occur. This has involved attempts to develop drought and famine early-warning systems, based on a range of environmental and socio-economic data such as rainfall levels, the condition of pasture, migration of people and livestock, level of crop storage and animal sales. However, such data are often unavailable and their interpretation can be time-consuming, requiring a deep understanding of the societies in question. By the time the data have been assessed the situation may already be serious.

In attempting to reduce vulnerability and improve food security it is important to investigate and treat each social group and environmental situation in its own right and build upon existing diversification, security and coping strategies. Governments, NGOs and others must work with local people to generate future development strategies which are sustainable. The concept of 'sustainable development' implies a need to manage natural and human resources in such a way that the quality of the resources in the long term is not jeopardized by the desire for short-term benefits. For example, if the intensive cultivation of cash crops is likely to lead to severe land degradation and soil erosion, then it will not be sustainable in the future and will result in the depletion of vital environmental resources.

Much has been learnt in recent years about the environment of tropical Africa. Slowly but surely we seem to be moving away from a viewpoint where African people are seen as passive victims of harsh and unreliable environments, towards a view that appreciates people–environment relationships and builds upon these in development strategies which work with the people and are sustainable in the long term. The main obstacle to such progress is likely to come from inappropriate and untimely political or military intervention.

Case study D

Coping with a marginal environment in Burkina Faso

The name Burkina Faso means 'land of noble people'. In the last two decades the people of Burkina (known as Burkinabe) have had to cope with the effects of two major droughts, and in certain parts of the country, where population densities are high and land use is intensive, serious land degradation has occurred requiring urgent remedial action. The mobilization of rural communities in generating greater environmental awareness and implementing a variety of conservation techniques has been impressive.

Before the change of name in 1984, the country was called Haute Volta (Upper Volta), a name chosen by the French colonialists who ruled the country until independence in August 1960. Like other countries in the savannah-sahel zone of West Africa, Burkina is very poor, with GNP per capita of $320 in 1989, life expectancy of 48 years and high infant mortality – 135 deaths per

Case study D (*continued*)

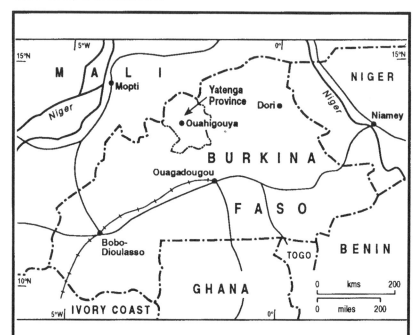

Figure 3.7 Burkina Faso

thousand live births. Its human development index in 1992, according to the United Nations Development Programme, was the fourth lowest in the world – only Guinea, Sierra Leone and Afghanistan had lower indexes.

Burkina, which is slightly larger than the UK, had an estimated population of 8.7 million in 1989, giving an average population density of about 32 per sq. km. However, in certain parts of the country rural densities are very much higher. More than 90 per cent of the population of Burkina live in rural areas, with 87 per cent of the labour force employed in agriculture. The capital city and administrative centre, Ouagadougou, is located in the middle of the country and has a population of about 450,000. The second major city and economic capital, Bobo-Dioulasso (with a population of 250,000), is located in a productive agricultural region close to the Ivory Coast border and

Case study D (*continued*)

on the railway that links Burkina with the sea at Abidjan.

French colonial rule created the main structure of the economy, with heavy emphasis on cash-cropping, notably cotton and groundnuts. The French regime was also responsible for initiating the large-scale migration of labour (mainly male) to neighbouring states, notably Ivory Coast and Ghana, where money for paying taxes could be earned in coffee and cocoa plantations or in the large towns. An estimated three million Burkinabe are today working in Ivory Coast, sending vital remittances back to their families in poor rural villages. The French colonial regime integrated Burkina into a system of regional trade and transport, with strong dependence on the more wealthy Ivory Coast. Burkina produced cotton and groundnuts, whilst manufactured and processed goods were imported from France. The lack of colonial investment in infrastructure is reflected in the fact that at independence there were no tarred roads in Burkina.

Yatenga, in north-western Burkina Faso, next to the border with Mali, is one of the country's poorest regions, with a harsh climate and short rainy season from June to September. Rainfall has decreased in the last twenty years from the long-term average of 720 mm per annum to 440 mm. Soils are poor, other than in a few river valleys, and much of the natural vegetation has been cleared for cultivation, causing rapid run-off and erosion during heavy downpours. Droughts have been recorded in Yatenga in 1832–9, 1879–84, 1907–13, 1925–6, 1929–34, 1940–2, 1966–73, and most recently 1979–84. In October 1984, after four bad harvests, the region had a gross annual cereal deficit of 149 kg per person.

Yatenga is the homeland of the Mossi people who make up 48 per cent of Burkina's population. The other main population group are the Fulani, predominantly pastoralists and comprising 10 per cent of the country's population. Yatenga is heavily populated, with densities as high as 70–100 people per sq. km in the central area and figures above 50 people per sq. km being common throughout the region. Pressure on the land, poverty of village life and attractions of wage labour elsewhere have caused many Mossi to leave Yatenga since the late-nineteenth century.

Case study D (*continued*)

Those Mossi who remain grow millet, sorghum and maize and in the colonial period the French insisted that all farmers should also grow cotton and groundnuts. Intensively farmed vegetable gardens, fertilized with kitchen waste, surround the villages. In many villages most of the land is under continuous cultivation with little possibility of expanding the agricultural area. Although tree cover has declined considerably, important economic trees such as the shea, dawadawa, kapok, tamarind and baobab are carefully tended and harvested. Unlike the Fulani, the Mossi traditionally kept few animals, but today cattle are seen as a form of investment and insurance against disaster. After the harvest, cattle graze crop stubble, adding valuable manure to the soil, which otherwise receives little fertilization, since chemical fertilizers are too expensive.

A number of recent projects in Yatenga have attempted to preserve and enhance environmental resources. For example, the Naam movement was founded by Bernard Ledea Ouedraogo, a teacher who moved into rural development work in the early 1960s. Ouedraogo's beliefs were, 'To make the village responsible for its own development, developing without destroying, starting from the peasant: what he is, what he knows, what he knows how to do, how he lives and what he wants.' The Naam movement, which started in 1967, derived its name from the traditional Mossi village organization which Ouedraogo chose as his starting point, a grouping of young men and women formed each rainy season to help with planting and harvesting. It is a traditional self-help cooperative movement, based on equality and with elected leaders. Wherever possible Naam activities use low-cost local tools and materials. In 1976 Ouedraogo set up an umbrella organization called 'Six S', to provide technical and financial help for Naam groups and to raise funds internationally for such items as medicines, cement, pumps and improved tools. By 1985 there were 1350 Naam groups and the idea had also spread to neighbouring countries.

The Naam group in the village of Somiaga built a pharmacy and a mill, dug several wells for drinking water, and established a tree

Case study D (*continued*)

nursery and a woodlot. A cereal bank was started in 1983 to improve food security in the village and to avoid major seasonal price fluctuations. Cereal stocks are bought cheaply at harvest time and stored until needed, thus avoiding expensive purchases in the late dry season or early rainy season when market cereal stocks are dwindling. A large dam has also been built, 180 metres long and 4 metres deep, to irrigate up to 60 hectares of rice, vegetables and fruit trees. It has raised the water level in local wells, made it easier to water tree seedlings and provided a useful source of fish. Up-stream from the dam smaller check dams were built across gullies to reduce the flow and collect the soil so the main dam would not silt up quickly.

The Agroforestry Project (PAF) is another example of a community-based venture to improve soil and water conservation in Yatenga. Started in 1979 as a tree-planting project, it has since diversified into promoting water harvesting and anti-erosion measures. Much of Yatenga has a gently undulating landscape and a key innovation was the simple plastic water tube level to determine contour lines for the construction of rock bunds. Rock bunds are a traditional method of water harvesting and are preferred to earth structures as they are not easily damaged by run-off and therefore require little maintenance. Also, as the bunds are permeable, crops planted in front of them benefit from run-off as well as those behind. Water retained by the bunds gradually permeates the soil, much of which is covered by a hard crust that would otherwise have prevented water infiltration. Trees and grass planted along the bunds help to further stabilize the soil and in time to produce barrier hedges.

Rock bund construction has enabled an expansion of the cultivated area through the reclamation of abandoned land. Since 1983 the number of hectares treated has doubled every year, reaching a total of 8000 hectares in over 400 villages in 1989. PAF has encouraged collective treatment of individual fields, and since many young men migrate in the dry season, women play a major role in rock bund construction. PAF provides training in the various techniques and supplies farmers with water tube levels,

Case study D (*continued*)

shovels, pickaxes, wheelbarrows and donkey carts. Oxfam has helped with the purchase of a tipper lorry to make stone gathering easier. Another technique known as zai is used in conjunction with the rock bunds. Zai are wide and deep planting holes which collect and concentrate run-off water and, when filled with manure and compost, greatly improve crop yields. The use of bunds and zai together produces crop yields averaging 972 kg per ha compared with an average yield on untreated plots of only 612 kg per ha. These simple techniques have provided greater food security by reducing the impact of low rainfall years on yields.

More recently PAF has broadened its outlook to promote an integrated approach to village land-use management. Village land-use management committees have been set up to decide what needs to be done to improve the whole area and then to coordinate various conservation strategies such as stone bunding, compost pits, the enclosure of sheep and goats in the homestead and village fodder plots. Communal land has been protected from overgrazing, tree nurseries started and a wide range of trees planted in the fields. PAF has had considerable success in enhancing the environment in Yatenga, due to its flexible, low-technology and community-based approach which has built upon a range of indigenous techniques. The Yatenga experience may well be usefully replicated elsewhere in tropical Africa.

Case study E

People and environment on the Jos Plateau, Nigeria

The traditional homeland of the Kofyar people is a territory of some 500 sq. km on the southern fringe of the Jos Plateau in central northern Nigeria. Kofyar economy and society are skilfully adapted to the rugged topography, and in recent years further adaptations have occurred, taking advantage of new economic opportunities and an easier lifestyle on the Benue plains below.

The Jos Plateau is a distinctive, mainly granitic upland area of about 8600 sq. km (80×110 km), with an average altitude of 1200

Case study E (*continued*)

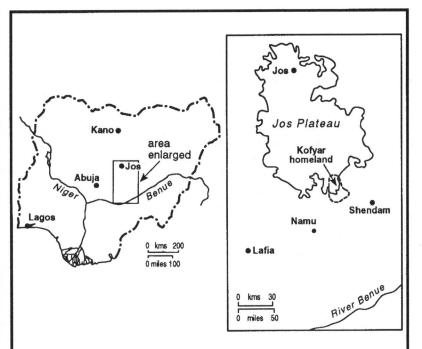

Figure 3.8 Jos Plateau showing the Kofyar homeland

metres, rising to 1350 metres on its southern margin, where there is a steep 300–600 metres escarpment overlooking the plains on the northern edge of the Benue valley. Old volcanic cones, volcanic boulders and lava flows are a feature of the area, on which immature and stony soils have developed in places with good fertility. Rainfall on the plateau is greater than on the plains, with most of the 1000–1500 mm coming between April and October, peaking in August. Temperatures range between 32 degrees centigrade in February and about 25 degrees during the rainy season, though night-time temperatures of under 10 degrees centigrade are not uncommon in the hills.

Kofyar country was settled as a defensive site in pre-colonial times when inter-tribal warfare was common, and includes a heavily dissected ridge 24 km long and 8 km wide, and an area of

Case study E (*continued*)

plainland at the base of the plateau escarpment. Following the British conquest in 1909 and the establishment of colonial rule, more peaceful times encouraged movement out of the hills to the plains, but this migration was not significant until the 1950s. At the 1963 census, the total Kofyar population was almost 73,000 with densities reaching 750 per sq. km, a very high figure for traditional African farming communities. Such dense settlement is supported in the hills by the intensive and continuous use of skilfully constructed terraced fields, where soil fertility must be carefully maintained to produce successive food crops.

Kofyar hill cultivation reflects an impressive understanding of the local environment's potential and problems, and gullying and erosion are rare. On steep valley sides terraces have been constructed with stone-built retaining walls usually placed at right angles to the slope. The terraces provide a level surface with deep soil for cultivation and help reduce run-off and erosion. On the fields of the homestead farms close to the villages, ridges of soil are made in a rectangular formation. Crops are planted on the ridges and their roots are watered from the water collected in the basin, an irrigation technique known as 'basin listing'. Groups of ridges are separated by drainage ditches to prevent waterlogging of the plots.

Unlike other African farming communities, the Kofyar do not experience a 'hungry season' before the harvest. They grow a wide range of grain, leguminous and tree crops which are harvested at different times of the year, ensuring a continuous supply of food. The food staples of sorghum, millet and cowpeas are usually intercropped, and other crops include acha or 'hungry rice', maize, yams, groundnuts, sweet potatoes and rice in moist areas. Kitchen gardens close to the houses provide herbs, peppers and tomatoes for sauces. Tree crops are also gathered, such as oil and fan palm, locust bean, paw paw and mango, further enriching the diet and sometimes being exchanged for other commodities. The key to the maintenance of such a productive farming system is the regular application of compost, wood ash and animal manure. During the growing season goats are staked in corrals close to the homestead and at night are locked into a hut. Their manure and trampled

Case study E (*continued*)

Plate 3.1 Village on the Jos Plateau, Nigeria. In this rugged terrain villages are nucleated and the small farms surrounding the settlements are intensively farmed, regularly manured and frequently terraced. The stone-built corral in the picture is used for keeping animals at night. Their manure is then spread on surrounding fields.

fodder provide a vital addition to the soil. On more distant bush-fields, which unlike the permanently cultivated homestead farms are only cultivated for six to nine years, Fulani herders are encouraged to walk their cattle over the land leaving valuable manure.

Technology is simple but effective on the small farms and consists of hoes, axes and sickles. Homestead farms average 0.62 hectares, whereas the less intensively cultivated and more distant bush farms are larger, averaging 1.2 ha. The average land requirement for a homestead is about 1.01 ha. The main problems facing the hill farming areas are overpopulation, limited farm size, poor services, difficult communications and access to markets. It was

Case study E (*continued*)

these factors which encouraged the Kofyar, quite spontaneously, to start migrating down the escarpment to establish farms on the Benue plains. Permanent colonization of the plains began in 1951 and increased rapidly in the 1960s. Initially there was regular movement between the plains and the hill farms, but in recent years many Kofyar have found it more convenient and profitable to live on the plains. Those remaining in the hills are often elderly or retired. A survey in 1984 showed that 74 per cent of those interviewed had migrated and over 50 per cent had transferred their sole residence to the plains, abandoning their homesteads and farms in the hills. By the mid-1980s some 25,000–30,000 Kofyar occupied the Benue Valley lowlands south of their original homeland.

In migrating to the plains the Kofyar have taken their knowledge and farming techniques with them, but certain significant adaptations have also been made. Plains farms are larger – an average of 4 ha compared with less than 1 ha on the homestead farms. Household size also increased in the 1960s from 4.17 people amongst non-migrants to 6.44 for migrants. By 1984 average household size had reached 8.38 people. Many migrant Kofyar are able to produce sizeable marketable surpluses. Itinerant merchants and lorry drivers visit their farms, many of which are within easy reach of the new surfaced road opened in 1982 linking the main production areas with key settlements and markets.

Migrant Kofyar household income has increased impressively in spite of little investment in new technology or hired labour. Traditional tools are still used and extra food production has been generated by farmers and their families working harder and longer hours in the fields. Kofyar farmers work an average of 1599 hours a year per adult worker, considerably more than the average 500–1000 hours recorded in similar food production systems elsewhere in Africa. This hard work is reflected in the way Kofyar farmers have extended the farming season, done away with slack periods and mobilized men and women equally in farm-related work.

The Kofyar migrants have adopted yams and rice as cash crops, and have also increased production and sale of traditional crops,

Case study E (*continued*)

such as millet and groundnuts. A new variety of yam from Iboland has been adopted which is of better quality and stores without shrinkage. Fewer goats are kept on the plains, but almost every Kofyar farmer now raises pigs, which were entirely unknown in the area as recently as 1960. Other innovations include growing cassava and bananas for the market and the application of chemical fertilizer. Money earned from crop sales is spent on clothes, taxes, bridewealth, bicycles, motor cars, school fees, medical care and chemical fertilizer.

The Kofyar example provides a good illustration of African farmers adapting to physical and economic environments. Opportunities were recognized and, quite spontaneously and without government or other agency involvement, farmers have colonized new land and by upgrading their labour input have developed profitable market-orientated food production businesses.

Key ideas

1 Tropical Africa has a wide range of different environments whose characteristic features are largely determined by latitude, altitude and rainfall.
2 Much has been written in recent years about desertification. It is not that desertification does not exist, but rather that its existence over space and time is much less than the popular image.
3 With inadequate data, changing environmental factors and the dynamics of human adaptation to environment, it is difficult to calculate the 'carrying capacity' of an area.
4 African food production systems frequently possess a high degree of structural resilience and adaptation to environment.
5 Various diseases, many of them associated with water, are widespread throughout tropical Africa, resulting in high mortality rates, particularly amongst children.

4
Rural Africa

Agriculture in the economies of Africa

Africa is still mainly a rural continent where most people are farmers, herders or fishermen. Though towns and industries are growing rapidly, by far the greatest proportion of Africa's people live in rural villages. For the foreseeable future, agricultural and rural development must be the main priorities if living standards are to rise and national productivity is to increase. Africa's political leaders seem to be aware of this and frequently make reference in their speeches and writings to the importance of rural development, but relatively little progress has actually been made in recent years. In fact during the 1980s Africa became associated with drought, desertification, famine and poverty.

In most African countries 70 per cent or more of the population work in agriculture (see Table 4.1 and Figure 4.2). Some countries, such as Rwanda and Burundi, have over 90 per cent of their populations working in agriculture. In very few African countries is the figure less than 50 per cent and this has scarcely changed in the past twenty years. Agriculture provides over 30 per cent of GDP in most tropical African countries, though in some, such as Malawi, Niger, Nigeria, Rwanda and Sudan, there was a marked decline in agriculture's contribution to GDP between 1965 and 1989. In some cases this reflects a sluggish agricultural sector, with a slowly growing industrial sector and a much more rapidly growing service sector – as in the cases of Niger and Sudan, for example. Those countries having only a small proportion of GDP generated by

Table 4.1 Tropical Africa: agricultural labour force and contribution of agriculture to GDP

	Percentage of labour force in agriculture		Percentage of GDP from agriculture	
	1965	1986–89	1965	1989
Angola	79.0	73.8	n/a	n/a
Benin	83.0	70.2	59	46
Burkina Faso	89.0	86.6	37	35
Burundi	94.0	92.9	55*	56*
Cameroon	86.0	74.0	33	27
Cape Verde	52.0*	44.0*	18*	15*
Central African Republic	88.0	83.7	46	42
Chad	92.0	83.2	42	36
Comoros	83.0*	80.0*	46*	34*
Congo	66.0	62.4	19	14
Djibouti	79.0*	77.0*	3*	2*
Equatorial Guinea	66.0*	57.0*	42*	59*
Ethiopia	86.0	79.8	58	42
Gabon	n/a	75.5	7*	10
Gambia	84.0*	81.0*	27*	27*
Ghana	61.0	59.3	44	49
Guinea	87.0	80.7	n/a	30
Guinea-Bissau	82.0*	79.0*	44*	46*
Ivory Coast	81.0	65.2	47	46
Kenya	86.0	81.0	35	31
Liberia	79.0	74.2	22*	37
Madagascar	85.0	80.9	25	31
Malawi	92.0	83.4	50	35
Mali	90.0	85.5	65	50
Mauritania	89.0	69.4	32	37
Mauritius	28.0*	23.0*	13*	13*
Mozambique	87.0	84.5	n/a	64
Niger	95.0	85.0	68	36
Nigeria	72.0	44.6	54	31
Reunion	18.0*	12.0*	6*	6*
Rwanda	94.0	92.8	75	37
Sao Tome and Principe	56.0*	53.0*	39*	24*
Senegal	83.0	80.6	25	22
Seychelles	11.0*	9.0*	7*	5*
Sierra Leone	78.0	69.6	34	46
Somalia	81.0	75.6	71	65
Sudan	82.0	63.4	54	33
Tanzania	92.0	85.6	46	66
Togo	78.0	64.3	45	33
Uganda	91.0	85.9	52	67
Zaire	82.0	71.5	22	30
Zambia	79.0	37.9	14	13
Zimbabwe	79.0	64.7	18	13

Sources:
1 Columns 1 and 2, United Nations Development Programme (1992) *Human Development Report.* Oxford: Oxford University Press, except those marked * which are for 1980 and 1989 figures taken from Commonwealth Secretariat (1991) *Basic Statistical Data on Selected Countries.* London: Commonwealth Secretariat.

2 Columns 3 and 4, United Nations Development Programme (1992) *Human Development Report,* Oxford: Oxford University Press, except those marked * which are for 1980 and 1988 figures taken from Commonwealth Secretariat (1991) *Basic Statistical Data on Selected Countries,* London: Commonwealth Secretariat.

3 n/a – data not available.

Plate 4.1 Enterprising farmer in Sierra Leone who has used his carpentry skills to make a threshing machine (right) and a winnowing machine (left) – good examples of intermediate technology, using indigenous skills and resources.

agriculture include Congo (14 per cent agriculture, 51 per cent industry and 35 per cent services) and Zimbabwe (13 per cent agriculture, 43 per cent industry and 44 per cent services). But the overall picture in Africa is one of a large proportion of the active population engaged in agriculture, which makes a significant contribution to national Gross Domestic Product.

Feeding Africa's people

With rapidly increasing population and periods of drought and recession in the wake of rising oil prices and fluctuating agricultural commodity prices, between 1971 and 1984 Africa experienced an average annual decline in per capita food output of 1.2 per cent (see Figure 4.1). In sharp contrast, comparable data for Asia show a rise in food output per capita of about 1.3 per cent per annum. In countries such as Angola, Mauritania, Mozambique and Somalia, the decline in food production was particularly significant. It is likely that the long-term decline in

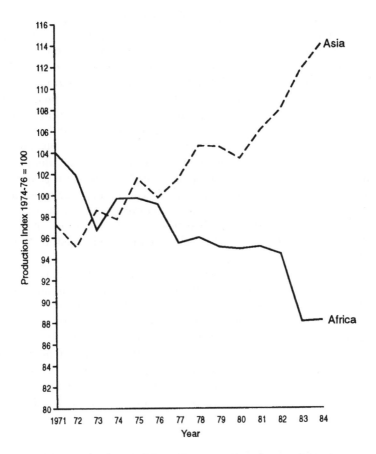

Figure 4.1 Food production per capita in Africa and Asia, 1971–84

African food production started in the 1960s, the decade when many countries gained their independence. The situation was exacerbated in some countries by major droughts in 1968–73 and the early 1980s and by political instability.

On 7th April 1985, the *Observer* newspaper suggested that:

In Africa there are 10 million people who have abandoned their homes and are looking for food, 20 or more countries critically affected by malnutrition and starvation, and 30 million lives in danger from famine.

Statistics of any sort in Africa should be treated cautiously, but in Ethiopia alone during 1984–5, famine affected up to eight million people and as many as one million may have died. The shock impact of the BBC film and Michael Buerk's commentary from Korem in Ethiopia on 23 October 1984 led to UN action and voluntary fund-raising on an unprecedented scale. Even under 'normal' circumstances, many Africans receive a daily calorie supply which is well below what they need, but during the two recent drought periods calorie intakes of people in certain African countries suffered considerably and large quantities of food aid were needed. Food aid figures for Ethiopia in the mid-1980s would have been very much higher if western governments had not been so wary of assisting the left-wing military regime which had overthrown Emperor Haile Selassie. However, the scale of western food aid to Ethiopia did actually increase as a result of the public outcry following the BBC film and the subsequent establishment of Band Aid in the UK.

There is no doubt that in certain African countries, agriculture has faced major problems, often exacerbated by civil war and political instability. But not all African countries face difficulties on the same scale as Ethiopia in the mid-1980s or Somalia in the early 1990s. Some countries are self-sufficient in basic foodstuffs and the levels of starvation portrayed in the media are fortunately not widespread in tropical Africa. Some African countries are successful exporters of agricultural produce such as coffee and pineapples from Ivory Coast and Kenya, tea from Kenya, cocoa from Ivory Coast, Ghana and Nigeria and cotton from Sudan.

Understanding traditional farming and pastoral systems

The agricultural sectors of African countries may be divided into traditional and modern sub-sectors. Typically, the modern sub-sector employs only a small proportion of the agricultural labour force, is more capital intensive, uses machinery, fertilizers and irrigation to produce crops mainly for export, and often depends on finance and expertise from overseas. The traditional farming sub-sector, on the other hand, employs the bulk of the agricultural labour force, is labour- rather than capital-intensive, uses techniques which have often changed very little in centuries, and is geared to producing food for domestic consumption. Although the traditional sub-sector employs many more people, it generally contributes less to government revenue than the profit-

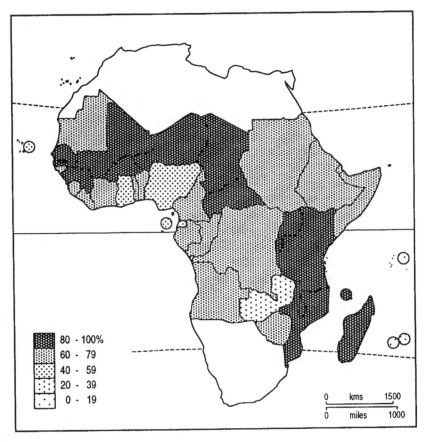

Figure 4.2 Percentage of labour force in agriculture, 1986–9

orientated modern sub-sector. But if governments are to improve the productivity of agriculture and raise living standards amongst the bulk of the population, it is the traditional farming and pastoral systems which will require more attention and help.

Textbooks frequently make the oversimplified distinction between *subsistence* and *commercial agriculture*, assuming that traditional farming systems are geared entirely to subsistence production and do not respond to market forces. Most farmers will, quite understandably, state that their main aim is to feed their families, but in fact some traditional farming systems produce large, though often variable, surpluses over and above household needs. Early observers of tropical

African farming systems frequently commented that farmers did not behave rationally, since profit maximization was not necessarily their main aim. Such a statement is misleading, since many African farmers, given good environmental factors and sufficient food for the household, will aim to sell as much surplus as they can and show a good awareness of market prices. The production and sale of such surpluses has been going on for centuries in some parts of Africa, and there exist long-established marketing systems with flourishing markets linked by trade routes. In pre-colonial times a variety of currencies such as cowrie shells and metal bars were in widespread use, but today these have been replaced by money. The fact that market places and trade routes have flourished over many centuries in parts of Africa suggests that production systems were geared not only to producing domestic needs, but also sizeable and regular surpluses for sale.

There has also been a tendency amongst outside observers, national governments and agricultural experts to view traditional African farming systems as primitive and environmentally destructive. However, recent studies have revealed that many of the techniques used by African farmers are in fact well adapted to the environment and might usefully be incorporated into future agricultural development programmes. African farmers know a great deal about their environment since their livelihoods depend on it. They show a good understanding of soil types, trees, plants and insect and animal pests and use a variety of different methods to preserve the environment and counter problems such as soil erosion. At times of stress such as drought, farmers use a wide range of techniques to maintain food supplies. African farming systems are remarkably resilient and when difficult periods are over most can recover quickly. It is important that those involved in agricultural and rural development in tropical Africa have an accurate understanding of the organization of food-production systems so that the most appropriate development measures can be selected for the benefit of both producer and consumer. It is particularly important to understand a farming system as a whole, monitoring it over a full annual cycle and considering the many interactions between human and physical elements (see Case study F).

Key features of African farming systems

There have been many attempts to classify tropical African farming systems, but because these systems are so complex and varied, no single

Plate 4.2 Farmer using traditional hoe west of Kano, Nigeria.

classification is entirely appropriate. Farming systems range at one extreme from extensive systems such as shifting cultivation, to intensively cultivated and permanent systems, such as horticulture, at the other. A useful classification would distinguish between farming systems depending on fallow periods and those with permanent cultivation. Permanent systems may in turn be sub-divided into small-scale and large-scale systems. The concept of a farm in Africa is very different from that in western Europe or North America. The term 'farm' usually means the same as 'plot' or 'field'. Whereas a British farmer would refer to the different fields making up the farm, an African farmer cultivating more than one plot would commonly refer to each plot as a separate farm. African farms are usually small, under five hectares being common, and farms and kitchen gardens of less than one hectare are not uncommon, particularly in areas of severe population pressure.

It is likely that all or most African farming systems once involved fallowing or resting part of the cropland for varying periods of time. In the absence of chemical fertilizers and sophisticated machinery, farmers rely on time and environmental processes to restore nutrients to the soil.

Fire was and still is used to clear plots in many parts of tropical Africa, and some farmers apply animal manure and kitchen waste to their farms to improve fertility and yields. Declining crop yields indicate that it is time to rest the land and cultivate elsewhere. In true shifting cultivation, which is now rare, this would involve movement of the household or entire community to another area of forest.

All farming systems are dynamic to a greater or lesser degree, and tropical African systems are no exception. They are affected by a wide range of factors such as population pressure, proximity to urban areas, and a variety of policies introduced by national governments and international development agencies. With increasing population growth and pressure on farmland in recent years, many farming systems have become more fixed, fences have been erected and fallow periods reduced considerably. With more intensive cultivation farmers are increasingly concerned to maintain fertility, and use manure, compost and also chemical fertilizers when they can afford them and they are available. The buying and selling of farmland, once unheard of in Africa, is now common in many places, particularly close to towns.

In general terms, the type of crops grown depends on the amount of rainfall, with root crops in the moister areas and grain crops in the drier areas. Maize, millet, sorghum, rice, yams and cassava are the staple food crops of tropical Africa, but not all of these are found in any one particular area. Rice, for example, is the staple crop of coastal West Africa from western Ivory Coast to Senegal, and is grown both on upland farms and in swamps. Yams are common in West Africa east of the Ivory Coast, whilst millet and sorghum are more common in the drier areas of interior West Africa, but also in Sudan and parts of East Africa. Two basic food crops, maize and cassava, are particularly widespread throughout tropical Africa, both having been introduced from the Americas. Maize is particularly important in Kenya, Zambia and Zimbabwe, though farmers elsewhere in tropical Africa also grow the crop. Cassava has increased in popularity in recent years, possibly more than any other crop, and has spread throughout Africa. It grows well on poor soils and gives good yields. Furthermore, it can be stored in the ground until needed, so the elaborate storage arrangements required for other major crops are unnecessary for cassava. The crop is valuable when other foods are in short supply, though it is starchy and not particularly nutritious. In addition to the staple foods, farmers also grow a wide range of other crops, particularly vegetables and fruits.

Farming activities in tropical Africa are closely related to the different

Plate 4.3 Women harvesting rice in The Gambia. Rice production is controlled by women. Each stalk of rice is cut one by one using a small pen-knife.

seasons. Most cultivation is undertaken during the rainy season with the harvest in the early part of the dry season. Farmers aim to maximise farm output and reduce risk by planting a number of different crops, quite often on the same plot. This practice of intercropping looks chaotic but is effective, giving good and reliable yields, whilst taking less from the soil than a single crop. The rainy season can be a difficult time and in some communities is referred to as the 'hungry season'. It is a time when the new crops are not yet ready, food from last year's harvest is dwindling and yet much work is needed on the farm tending the growing crops. It is at this time that cassava or some early corn cobs can help to fill empty stomachs.

Another feature of farming systems in tropical Africa is the low level of technology and the crucial importance of human labour. Basic farm technology has changed very little over many years, though some farmers have adapted to particular situations, for example dry-season irrigated farming in northern Nigeria where many farmers use simple *shadufs* to raise water from wells. With only simple technology, crop

Plate 4.4 Hausa woman inside her fenced compound with her family's storage baskets, east of Kano, Nigeria. The picture was taken towards the end of the dry season when stored food stocks are often running low.

yields are heavily dependent on the amount of labour in the farm household. Households with a large proportion of old or young members who are unable to work on the farm may have difficulty producing enough food and may be forced to buy or borrow food in the hungry season. Women play an important role in producing food in tropical Africa, and it is estimated that well over 70 per cent of Africa's food is produced by women. Many of the most demanding farm jobs, such as hoeing, weeding and harvesting are done by women, in addition to taking care of young children, collecting firewood and water and processing and cooking food. The labour provided by children is also important, in jobs such as bird-scaring, which, if not done by them, would probably be yet another chore for the women.

Tropical African farming systems could certainly be improved with better technology, but with the present living standards of most farmers, such technology is well beyond their means. What must be appreciated is that given the available resources, many of these farming systems have reasonable levels of productivity and are well adapted to the environments in which they operate.

Pastoralism

The rearing of livestock, often over large areas, is found mainly on the savannah grasslands of East and West Africa where rainfall may be less than 500 mm per year and where the quality of pasture is often poor. Pastoralists have a marginal existence in harsh environments, and those who inhabit the drier regions of the West African Sahel and the Eastern Horn of Africa suffered considerably during the droughts of 1968–74 and 1979–84.

Pastoral systems, like cultivation systems, have often been misunderstood. Pastoralists are frequently accused of being unhealthy and uneducated and keeping livestock purely for prestige purposes. Furthermore, their migrations, which are an important feature of most pastoral systems, are criticized by governments for disregarding international boundaries and making census-taking and tax-collecting difficult. Many governments would prefer pastoralists to settle and take up commercial ranching, supplying urban populations with reliable supplies of meat and milk.

The frequency of movement and distances covered vary considerably among pastoral groups. Among the Fulani of northern Nigeria, for example, the Bororo are true nomads, but are few in number compared with other semi-nomadic and settled Fulani groups (see Case study G). Very few pastoralists depend entirely on their animals for a living and most nowadays grow crops to feed their families. The Wagogo of Tanzania, for example, regard themselves first and foremost as livestock herders, but they also undertake a considerable amount of cultivation. Similarly, amongst the Karimojong of north-eastern Uganda, the women live in homesteads in well-watered areas and grow crops, whilst young men take herds of cattle, sheep and goats in search of pasture. Pastoral groups may develop links with neighbouring farming communities, as is the case in The Gambia and northern Nigeria, where Fulani pastoralists graze their animals on farm stubble after harvest, adding valuable manure to the soil. However, disputes occasionally occur if livestock disturb growing crops. The exchange of dairy products for grain is also common, and complex systems of loaning animals exist amongst pastoral groups, sometimes for payment and sometimes as a free gesture in return for help at some future time.

Pastoralists commonly keep a number of different types of animals, because not all will equally suffer during drought periods, which are common in the areas they inhabit. Camels, cattle, sheep and goats have different food requirements and varying degrees of resistance to drought and disease. Camels and goats are particularly resilient and it is not uncommon in the Sahel to see goats grazing several metres above the ground in thorn bushes. Pastoralists have a detailed knowledge of the environments in which they live, and in difficult places with limited resources, human birth rates are usually lower than amongst settled farmers and ranchers. Some pastoralists also control the breeding of their animals so that births occur when pasture and water availability are best. Female animals are particularly important and are seen as the main capital stock from which herds may be expanded after drought periods. The Tuareg of northern Mali sometimes sell young male animals to buy grain from farmers. Their male camels and donkeys are used for transport, while young male goats are slaughtered shortly after birth to release milk for human consumption. Female animals, however, are usually only sold as a last resort, such as during the drought of the early 1980s, when many Tuareg lost all their stock and had no females to rebuild herds.

Over-grazing can be a problem when animals are owned by family groups, yet pasture is managed communally. There is often a tendency for individual families to take more than their share of pasture, and sometimes it is necessary to close particular areas for grazing to allow pasture to recover. The Borana of Ethiopia manage their pasture and water carefully, since these are the fundamental resources for their existence. Wells are the focal points of Borana social life and the most vital ecological resource and as such are controlled and maintained by well councils.

Living in marginal areas with unpredictable climates, pastoralists have developed flexible lifestyles to respond quickly to periods of drought and famine. Exchanges of livestock and produce, changes in migratory routes and the use of wild foods for human and animal consumption are all common adaptive strategies used in difficult times. Whilst some would advocate the settlement of pastoralists on ranches, others would argue that pastoral areas could be used for little else, so it is better to encourage pastoralism by raising living standards and reducing vulnerability.

Plate 4.5 Second hand clothes market, Banjul, The Gambia.

Rural marketing

Rural markets are a common and colourful feature of many parts of tropical Africa. Their presence and permanence indicates that there are economic incentives to produce surpluses over and above household subsistence requirements. African markets perform a number of different functions such as:

- centres of collection and exchange of local foodstuffs, livestock and craft goods;
- provision of services, such as cooked meals, tailoring and repair of shoes, watches and bicycles;
- distribution points for goods coming into the area, either from other areas in the country or from abroad;
- collection points for goods that will be exported from the local region.

A single market may fulfil all of these roles, as well as perform important social functions. In remote rural areas without access to newspapers and other media, markets act as meeting places where people can catch up on news and gossip. Although media and fast transport may be lacking, it is surprising how quickly news and information about prices travel through rural Africa. Some parts of tropical Africa have a longer tradition of market-place trading than others. Some West African countries, for example, have market systems dating from long before the colonial period. Markets in Uganda, however, seem to be much more recent phenomena, associated with the sale of manufactured goods and service provision. In the non-Muslim parts of West Africa, such as Ashanti in Ghana, or Yorubaland in Nigeria, many of the traders are women. In Muslim areas, such as northern Nigeria and coastal East Africa, men are the dominant traders.

A major distinction can be made between permanent markets and temporary or periodic markets. Periodicity of occurrence is a feature of well over three-quarters of all African markets. One of the most intricate systems of periodic markets is to be found in Yorubaland, southwestern Nigeria. Five main types of market exist according to location and timing: urban daily markets in the largest towns, urban night markets, rural daily markets for the sale of fresh produce and meat, rural periodic day markets and rural periodic night markets. Rural periodic day markets are particularly important, being located about seven miles apart, with no village more than five miles away from a market. By taking place at different sites on different days, a cycle of periodic markets can draw on a larger population of buyers and sellers than a single fixed market. Yoruba periodic markets are arranged in ring systems, normally in four- or eight-day cycles, and successive markets are not normally adjacent. A study of the Akinyele eight-day ring found that most women visited two out of eight markets in the cycle, and in any one day some 10,000 people were on the move, representing 35 per cent of the total population.

Most market-places in north-western Cameroon also operate on eight-day cycles, though with increasing commercialization daily markets are becoming more common. Most rural consumers visit only one large market each week to buy commodities such as palm oil, salt, dried fish or meat and occasionally a small supply of rice or some other commodity not normally grown on their farm. Until recently, Sierra Leone had no major periodic markets, possibly due to the lack of a basis for inter-regional trade, since goods produced in one area of the country are simliar to goods produced in other areas. However, since the 1960s some large rural periodic markets have developed in the Eastern Province, initiated by local chiefs and satisfying the increasing demand

Plate 4.6 Rope market at Dambatta, northern Nigeria. Such large markets, which are usually held weekly, provide an important focus for the exchange of locally produced materials and a meeting place for the exchange of news and gossip.

for foodstuffs in the growing urban centres of the diamond mining region. Much of the produce is transported over long distances to these weekly markets, but local farmers are also benefiting from new trading opportunities and from selling increasing amounts of fruit and other foodstuffs.

Certain aspects of rural marketing may be under the control of the government. In some tropical African countries rural producers may sell cash crops to agents or buyers from state-run marketing boards, although under recent structural adjustment programmes some of these boards have been dismantled. Commonly, farmers deliver their produce to traders or buying points, though sometimes buyers will purchase crops actually on the farm. These boards were established after the Second World War in countries such as Nigeria, where they were responsible for buying cocoa, palm produce, groundnuts and cotton, and in Ghana where they purchased cocoa. Marketing boards aim to stabilize producer prices and foreign exchange earnings and reduce inter-seasonal price movements. When world prices are above producer prices, the surplus earned by the board is put into a stabilization reserve to be used to finance the deficits of other seasons when world markets fall below producer prices. In theory, the average producer price over a period of years should roughly equal the average net price realized in the export market. However, marketing boards have been criticized for not successfully stabilizing prices from season to season and for failing to stabilize incomes. It is also argued that the boards have a generally depressive effect on the economy by paying low producer prices, whilst accumulating large reserves.

Good marketing facilities with attractive producer prices are essential if farmers are to be encouraged to produce more food and cash crops. Such facilities need to be combined with an upgrading of transport and refrigeration for perishable goods.

Rural industries

Although most people in rural areas are engaged in farming, different types of non-farm work have long been a feature of rural communities and can make a valuable contribution to household income, particularly in areas with a long dry season where cultivation is not possible for perhaps half the year. Fishing, weaving, metalworking, carpentry, pottery making, trading and even music-making are typical of the non-farming activities to be found in rural villages, using skills which have often been handed down from generation to generation. It is important in any future rural development strategy that such skills are recognized and strengthened in an effort to diversify the economies of rural communities.

Rural enterprises are generally family-owned and rely largely on indigenous resources for finance, raw materials and labour skills. Blacksmiths, for example, often make their own tools, and their raw materials come from discarded metal objects such as car springs. Their skills may be harnessed to make improved farm technology such as donkey carts and ox-ploughs. Many non-farm activities are carried out in the seclusion of the household, so data are difficult to obtain. However, it is estimated that the non-farm sector provides primary employment for between 6 and 26 per cent of Africa's rural labour force, and in countries where data are available, the total numbers employed in rural manufacturing exceed those employed in urban manufacturing. In addition to providing primary employment, there are

Plate 4.7 Blacksmiths working metal from abandoned cars in northern Nigeria.

many rural people who get involved in non-farm work on a more casual basis. Non-farm activities provide important employment for disadvantaged groups such as women and the landless. The proportion of women in non-agricultural employment varies considerably from about 12 per cent in Malawi to possibly 65 per cent or more in Ghana and Zambia. In countries such as Ghana, The Gambia and Senegal, for example, women control most of the marine fish marketing and processing. In Tanzania, rural women have multiple income generating activities, with beer making, cooking and food selling being most important. Rural non-farm activities can play an important role in reducing poverty, since wages outside farming are generally higher. In a study of three villages in Kano State, Nigeria the increasing proportion of income generated off-farm was a major factor in village-level inequality. Those people with small farms are likely to earn a greater proportion of their income from non-farm sources than those with larger farms. In Kenya it has been suggested that the growth of non-farm incomes has actually reduced rural–urban migration.

A dynamic rural non-farm sector can contribute to raising farm productivity and vice versa. Rural non-farm activities have been neglected by African governments, and their future development could be accelerated with the provision of roads, telecommunications, electricity and credit institutions geared to supplying finance to people with few assets. Schools should also be encouraged to teach craft skills and simple enterprise management techniques.

Cash crops and food crops

There is much current debate about the merits and problems of growing crops for sale and export rather than for domestic consumption. Although any crop which is traded for money might be called a 'cash crop', the term is usually used to refer to major export crops such as coffee, cocoa, tea, sisal and groundnuts. Large-scale production of these crops in tropical Africa started in the colonial period, when the European powers were looking for places to produce cheap raw materials for developing food and manufacturing industries. Export crops are basically produced in two ways, on smallholder farms and on large estates or plantations. Plantations are changing, but their typical features include:

- expatriate ownership and management, often by a multinational company, with production of a limited range of crops geared to the export market and some processing of the product being undertaken on the plantation;
- use of labour-intensive production methods, and a relationship between management and labour force which is hierarchical and characterized by inequalities and restrictions;
- they are 'total institutions', often self-contained communities, with workers living on the plantations and being heavily dependent on them for their livelihood.

The establishment of expatriate-owned plantations was discouraged in British West Africa. However, the French had plantations growing coffee, cocoa and bananas in Ivory Coast and bananas in Guinea, the Americans grew rubber in Liberia and the Germans, as early as 1884, set up rubber, oil palm and banana plantations in Cameroon. Elsewhere, the Belgians developed rubber, oil palm and coffee plantations in Zaire. In eastern and southern Africa, white settler farmers established estates in countries such as Kenya and Zimbabwe growing tea, coffee, pyrethrum, tobacco, cotton and sugar. Since independence, some countries have witnessed a retreat of foreign companies from estate ownership, and in certain cases plantations and estates have been nationalized, though in Tanzania this was associated with a decline in managerial efficiency, profitability and investment levels. Despite past criticisms, there is now more support for plantations among Third World governments, and the movement to nationalize plantations is no longer quite so fashionable. Kenya has a valuable plantation sector co-existing with smallholder agriculture, and evidence suggests that the poorest rural people may actually be better off within a plantation environment than if they remain as landless labourers in some of Kenya's poorest districts. In addition to providing employment, some large Kenyan plantations also have health-care facilities which are superior to those in surrounding areas.

A major concern over cash crop cultivation on smallholder farms relates to problems of food security and the nutritional status of households which may be neglecting food production to cultivate crops for the market. There are many examples of cash crop-growing communities with poorer nutritional status than more subsistence-oriented communities in similar areas. Cash crops are often the main responsibility of men, whilst women continue to grow the staple foods. In The Gambia, for example, women grow rice, whilst men are more concerned with the groundnut crop and upland cereals such as sorghum, millet and maize. The neglect by men of food crops in favour of cash crops can place a considerable burden on women in trying to grow enough food to meet subsistence needs. In some communities the rewards of cash crop production are kept by the men, and since the women are fully preoccupied with feeding the family, they have little opportunity to earn any additional income. There is a need for more detailed studies on the relationships between cash crop production and household nutrition.

Large- or small-scale development schemes?

A further area of debate concerns the most appropriate size, style and focus of rural development projects in tropical Africa. After the Second World War, in the last years of colonialism, the European powers were keen to establish schemes in Africa to produce much-needed agricultural resources. Reflecting the wartime mentality, a number of vast agricultural projects were launched with dismal results. Possibly the most infamous of these schemes was the ill-fated East African

Groundnut Scheme, designed to satisfy an urgent need for edible fats and oils in post-war Britain. A mission was sent out to East Africa in 1947 and after a rapid nine-week survey concluded that a vast area of 1.3 million hectares could be cleared and planted with groundnuts, mainly in Tanzania (then Tanganyika), but some also in Zambia (Northern Rhodesia) and Kenya. The scheme was to be implemented between 1947 and 1952 and involve 1000 Europeans and 60,000 Africans in land-clearing and cultivation. Much ex-army personnel and equipment were employed, including converted Sherman tanks used as tractors. The scheme proved to be a disaster and caused a political storm in Britain. The area was very dry, there was little available data on rainfall and soil characteristics and insect infestation was a problem. No pilot scheme was undertaken and the machinery, which suffered from breakdowns and fuel shortages, proved totally inappropriate for clearing the vegetation. The port of Dar es Salaam had great difficulty coping with the fourfold increase in tonnage that occurred. The scheme represents a classic example of failure due to inadequate evaluation of natural and human resources. No one thought to talk to the indigenous Wagogo pastoralists about environmental conditions. They were merely recruited on to the scheme as servants and labourers. By the time the scheme was written off in 1951, £36 million of British taxpayers' money had been spent.

Many other experiences of the failure of large-scale schemes could be recounted. What is regrettable is that some governments do not seem to have learnt any lessons from these failures and such projects are still being implemented in tropical Africa. Nigeria, for example, is still pursuing a 'big is best' approach in its development of a series of costly large-scale irrigation schemes in the north of the country, in some cases with considerable local resistance. In 1980 farmers living on the site of the proposed Bakalori project near Sokoto in north-western Nigeria rioted because of dissatisfaction with compensation for their farmlands and homes flooded under a reservoir 19km long. Anti-riot police were called in and opened fire, killing fourteen farmers and wounding fifteen. Results of these projects have often been very disappointing, with poor yields due to unpredictable supplies of inputs (water, seeds, fertilizers, pesticides, mechanized tillage) and poor management. Farmer incomes are low and poorer farmers on the Kano River Project in Nigeria have sold their farms, in some cases seeking work from larger, land-purchasing neighbours who may be wealthy urban-based, absentee politicians and businessmen. Large dams built across rivers can have a profound effect on downstream areas, preventing annual flooding and reducing fish catches, crop production and grazing reserves. A serious flaw in northern Nigeria's strategy of large scale irrigation schemes is that they have not been integrated with a coherent system of river basin planning.

The weaknesses of such large-scale schemes are slowly, but belatedly, being recognized, and governments in tropical Africa are now showing more interest in small-scale, low-technology development projects working with local communities. There is much unrecognized potential, for example, in developing small-scale irrigation schemes related to the moist *fadama* areas of northern Nigeria or the small valley wetlands known as *dambos* in Zimbabwe and other parts of southern and eastern Africa. In both cases these areas produce large quantities of vegetables and other crops with minimal capital inputs and with few, if any, adverse environmental consequences. The careful development of such resources could bring significant benefits to rural people with limited financial and environmental cost.

Future rural development strategies

The poor performance of the rural sector has been a major contributor to the crisis in tropical Africa. Despite thousands of development projects and massive external support in the shape of overseas aid and personnel, there seems to have been little real improvement in living standards for most Africans. Perhaps the nature of rural development strategies since the Second World War should take at least some of the blame. The focus of so-called 'development' projects has often been on increasing growth and raising productivity, rather than concentrating on generating true development in the shape of meeting basic needs and raising living standards. Many projects have been large, heavy on technology and dictatorial in the way they have related to those being 'developed'. Not surprisingly many rural people have felt alienated.

There is a small glimmer of hope in recent approaches to African rural development. It has at last been realized that before change is introduced, existing systems, with their natural and human resources, must be fully appreciated. The importance of households as the key production, consumption and decision-making units must be recognized and household dynamics must be understood. On the basis of this information, the aspirations and knowledge of rural people can then be incorporated into development schemes. Projects will vary in character from one community to another, but the broad focus should be on strengthening smallholder production, reducing vulnerability, increasing food security and raising living standards. Attention must also be directed towards improving health and education standards in rural communities. Hopefully such projects will diversify and enhance the productive base of rural areas, making them more attractive places in which to live and work and thus helping to reduce the drift to towns. Large projects will be replaced by small, community-based and more democratic schemes. All this, however, requires a genuine commitment on the part of African governments to moving rural and agricultural issues to the top of their political and financial agendas. Whether these aims will actually be achieved remains doubtful given the urban bias of government policies and the powerful voice of the urban people.

Case study F

Indigenous agricultural systems in Sierra Leone

Sierra Leone, a small (72,335 sq. km) West African republic which was a British colony until independence in 1961, has a tropical climate with an annual mean temperature of 26.7 degrees celsius. Annual rainfall reaches 6700 mm near Freetown, the coastal capital, but decreases inland and eastwards to 2200 mm. Rain falls mainly between May and October, with a dry season from November to April. The cultivation cycles of traditional non-irrigated farming systems are closely related to these seasons, with the main cultivation period corresponding with the rainy season.

Kono District in north-eastern Sierra Leone is an area of plateaux and hills separated by river valleys. Primary forest is only found in protected reserves, much of the present vegetation being secondary bush, the result of repeated clearance and cultivation by farmers. Further north, with declining rainfall, savannah grassland interspersed with woodland is common. Soils are generally light, lateritic and naturally infertile, though they receive a plentiful supply of nutrients from decaying vegetation. The most fertile soils are along river valleys where there are thick deposits of alluvium and inland swamps.

Land in Kono is managed by chiefs on behalf of their communities and individual farms average 1.25 hectares. Indigenous Kono

Case study F (*continued*)

farmers can clear land where they wish, provided no other person has already started work. Outsiders, however, are required to approach the Paramount Chief, who allocates land usually for one cultivation cycle in return for a portion of the harvest as payment. Buying and selling of land is rare away from the main towns of Kono District. The upland rice farm is the most important food production system, and though some farmers have swamp farms, kitchen gardens and cash crop gardens, the rice farm is the focus of household and community economic and social activity. A rotational bush fallowing system is used, with periods of one or two years' cultivation separated by long fallow periods. Settlements are fixed, with most farmers walking from the village to their farms each day during the cultivation season.

The location of a farm may be chosen several years ahead, with farmers taking simple soil tests and examining other natural indicators of fertility. Clearing normally starts in the latter part of the dry season during January, by which time much of the vegetation is dry and brown. The undergrowth is cleared using a cutlass and then trees are felled with an axe at about a metre from the ground, leaving the stump and roots intact. These are difficult to remove with available technology, but tree roots also help to bind the soil together and prevent erosion. When the farm is abandoned, trees and other vegetation quickly grow back, returning valuable nutrients to the soil. Certain economic trees, such as the oil palm and fruit trees, are not felled. The oil palm is fire-resistant and provides oil for cooking and an intoxicating liquid known as palm wine.

Deciding when to burn the farm can be difficult, but it usually takes place in March or early April when the first storms occur. If burning is done too soon then the debris may not be thoroughly dry, but if it is left too late, the rain may extinguish the fire, necessitating the laborious re-gathering and re-firing of the debris. The whole household helps to burn the farm, ensuring that the fire neither destroys economic trees nor strays on to neighbouring farms. After burning, timbers are used to build a barn, which serves as the main shelter during the cultivation

Case study F (*continued*)

season and may also be used for storing the harvest.

When it is raining regularly, in late April or early May, the seeds are sown. A technique known as 'intercropping' is used, in which up to fifteen or more different crop types may be grown on the same farm. In addition to rice, it is not uncommon to see yams, beans, tomatoes, okra, benniseed, maize, cucumber, cotton and other crops. This technique was widely criticized during the colonial period, since it was thought that yields of individual crops were lower than in monocropped fields. However, in recent years the many virtues of intercropping have at last been recognized. For example, different crops make different demands of the soil and cover the ground, protecting the soil from raindrop impact and possible erosion. Total crop yields may actually be greater than on monocropped fields and intercropping provides the household with a varied diet.

Rice and other seeds are broadcast and quickly covered by men using long-handled hoes. Other crops, such as maize and yams, may have been planted earlier. Maize, for example, is often one of the earliest crops to be harvested, in July or August, and can be a welcome relief during the pre-harvest 'hungry season'. With more frequent rainfall, the crops grow quickly and regular weeding is undertaken by women using short-handled hoes, this work being regarded as too degrading for men. It is back-breaking, time-consuming work which women do in addition to preparing food, tending vegetable gardens, collecting water and firewood, looking after young children and selling surplus produce in the market. By late June and July the maturing crops have to be protected from pests, notably the 'cutting grass' or cane rat, a large beaver-like rodent which, in the space of a single night, can flatten and eat large areas of crops. Fencing and trap-making, undertaken by the men, is difficult and unpleasant work since it often has to be done in the rain. August is the wettest month and often a time of food shortage and illness, which may lead to a smaller harvest if farm work is neglected.

Rainfall decreases in September and, as the rice crop swells and ripens, bird-scaring platforms are erected in the fields. Children

Case study F (*continued*)

spend many hours keeping birds away from the crops and they compete with each other in devising ingenious bird-scaring strategies. The rice harvest between mid-September and November is the big event of the year and the entire household helps. Labour supply is the main limiting factor on Kono farms and although a large household with many adults may have the necessary labour force, it also needs to produce a lot of food. Households with a high proportion of very young or old members are often constrained in their farming unless they have access to voluntary help from outside the household or can afford to hire labour. To avoid labour 'bottlenecks' at certain times of the year, farmers may subdivide their land and stagger the operations so that, for example, the entire farm does not have to be weeded or harvested at the same time.

Rice is cut stem by stem using a small knife and gathered into bundles to dry in the sun, after which it is stored in the barn or carried to the village. Most of the rice crop is used for household consumption, but some may be allocated to settle debts or for festivals and ceremonies, such as the initiation of teenage children into Kono secret societies. A good harvest may lead to rice and other crops being sold in the market, though there is nothing more humiliating than selling produce for a low price at harvest time, only to find that by the next rainy season the family is short of food and expensive purchases may be unavoidable.

After the harvest men often engage in various crafts, build and repair houses or test their luck at some alluvial diamond mining. Women continue to visit the farm to gather the remaining vegetables and may even take over a small portion of the farm as a cassava garden for a year or two. But no sooner is the harvest completed than January arrives and a new farming cycle on a different farm commences.

The Kono upland rice farm is carefully adapted to environment, technology and labour. With low population density, fallow periods of 8–15 years are possible and there is no significant shortage of land. However, greater population pressure could lead to land competition and the need for farming systems to adapt

Case study F (*continued*)

> further. This could involve the buying and selling of land, fixed farm boundaries and the use of manure and compost to maintain soil fertility.

Case study G

Pastoralists and pressures in northern Nigeria

The savannah lands of northern Nigeria have been ranged by Fulani pastoralists for centuries. They migrate with herds of cattle, sheep and goats in search of pasture and water. In recent years, however, the Fulani have had to adapt to new pressures which threaten their pastoral lifestyle.

The Fulani probably arrived in Nigeria in the early fourteenth century after migrating eastwards from what is now Senegal. They were gradually accepted by the resident Hausa population and during the jihad or Holy War from 1804, strengthened their own position as well as that of Islam. The Hausa kings were overthrown and replaced by Fulani Emirs. The new ruling class controlled grazing areas, cattle tracks and watering places, and a growing interdependence developed between pastoralists and sedentary farmers, a relationship which has often been called 'symbiotic', a mutually beneficial partnership between different groups.

This symbiotic relationship continues today, and a visit to the rich agricultural area around the historic city of Kano in February or March at the height of the dry season would reveal herds of animals wandering across land which during the rainy season (April–September) is covered with millet, maize, sorghum, beans and groundnuts. Livestock graze on any vegetation they can find, depositing manure, for which the farmers may sometimes pay the herdsmen. This is the key to a highly productive farming system, where fallow periods are virtually non-existent and the same fields are cultivated year after year.

Not all Fulani are migratory; a large number of them are now settled and this process seems to be accelerating with modern day

Case study G (*continued*)

pressures. Four major groups of Fulani may be identified according to their level of settlement, migration patterns and the amount of land under cultivation. The Bororo are true nomads, shifting camp regularly and moving south in the dry season and north in the rainy season. Semi-nomadic Fulani have permanent homesteads with adjacent farms where a variety of crops are grown. The elders stay in the homesteads for most of the year, to be joined by the returning young herdsmen at the start of the rainy season when the fields must be prepared for planting. Semi-settled Fulani regard cropping and livestock-rearing as equally important. They have fixed settlements and migrate over shorter distances, mainly during January and February. A fourth group, settled stockowners, graze or corral their herds close to the villages and young men take animals out daily into the surrounding area. This group increasingly supplements its income by looking after animals owned by wealthy town-based people who invest in livestock as symbols of wealth and status. Some stockholders are also engaged in the meat trade, supplying livestock to local markets and arranging transportation of animals to the large urban markets of the south. All four groups of Fulani show a common attachment to their animals and are held in esteem for their skills at rearing and managing livestock.

The Fulani calendar is closely related to climate, quality of grazing, water resources and the prevalence of disease, and can be divided into six eco-periods (see Figure 4.3). Pastoralists and cultivators anxiously await the first rain of the *seeto* season. Until then, hot humid conditions increase contagious livestock diseases and mortality rates rise. Young herdsmen must be prepared to move frequently in search of areas where rain has fallen, but as the rainfall becomes more regular in May they start to move back to their homesteads. Grass usually grows quickly in June, *seeto luginni*, and rain may fall every day. During *ndungu*, the period of heaviest rainfall, grazing quality improves, milk yields increase and female animals give birth. Harvest time from late September to early December (*yamde*), is when herdsmen plan their routes for the coming dry season. After about three months when the

Case study G (*continued*)

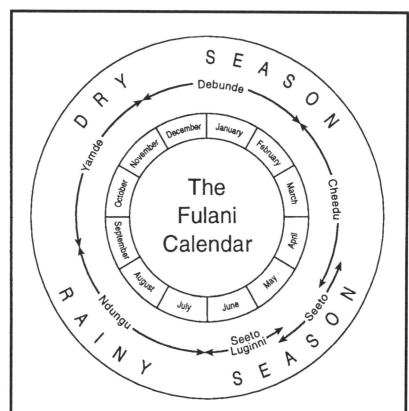

Figure 4.3 The Fulani calendar

cold and dusty Harmattan wind blows southwards from the Sahara (*debunde*), southerly winds herald the start of *cheedu*, the hot dry season from late February. Morale is low at this time, milk yields decline and if the dry season is protracted as it was from 1968 to 1973 and again in the early 1980s, animals may suffer and die.

The provision of adequate supplies of food and water for their animals is the main concern of Fulani herdsmen. After the rainy season, in late September and October, grass grows abundantly, but then slows down and loses much of its nutritive value as the dry season progresses. It is then that other sources of fodder must be

Case study G (*continued*)

found. Crop residues such as sorghum stalks, left in the fields after harvest may be eaten by the animals, though farmers may guard these carefully for fuel or house and fence construction. Low-lying depressions and river valleys, known in northern Nigeria as *fadama* land, are important dry-season sources of pasture and water, though herdsmen must prevent their animals from damaging crops. Leaves, flowers and fruits of the many large trees in this 'farmed parkland' of the northern savannah are also important sources of fodder.

Although the Fulani lifestyle has evolved over many centuries to encompass a detailed knowledge of environment, rainfall patterns, grazing resources, water availability, pests and diseases, it is a hard and uncertain existence; one which is becoming even more problematic as the Fulani face rapid changes in the area around Kano. Population pressure is increasing, with densities of more than 200 people per square kilometre making the 'Kano close-settled zone' (KCSZ) amongst the most densely settled land in rural Nigeria and indeed in tropical Africa. The efficiency of the farming system is maintained by intercropping a variety of grains and legumes on regularly manured land. More than 85 per cent of the total area is now cultivated and land values and land sales are increasing. On the outskirts of Kano every scrap of cultivable land is used to grow crops, and where there is well-watered *fadama* land, vegetables are intensively cultivated and transported by lorry to the city's markets.

With this intensive use of land, the Fulani have increasing difficulty finding pasture and water. Farmers are concerned about animals damaging their crops and often fence off *fadama* areas. The symbiotic relationship between farmers and herders is not always as harmonious as it is portrayed, and tensions frequently occur. With increasing population pressure and competition for farmland in the KCSZ, the pastoral free-range concept of land no longer meets with the approval it once enjoyed. Drought and disease have further troubled the Fulani in recent years. The much publicized drought of 1968–73 and the more recent droughts of the early 1980s have depleted herds and created shortages of food and

Case study G (*continued*)

water. The cattle disease rinderpest has killed many animals and a hard-pressed veterinary service has been unable to maintain an effective large-scale immunization programme.

Fulani pastoralists have also been affected by the construction of dams and irrigation schemes, which since the mid-1970s have transformed the landscape of northern Nigeria. Completed in 1976, the Tiga Dam, with a capacity of 1963 million cubic metres of water, irrigates over 60,000 hectares of land on the Kano River Project, where crops such as wheat, rice and tomatoes are grown. East of Kano in the Hadejia valley, a further irrigation scheme covers 25,000 hectares associated with a barrage across the river near Hadejia. Fulani interviewed in 1984 spoke of difficult times with declining herd sizes due to severe drought and rinderpest. Whereas the Tiga Dam and Kano River Project had made water more plentiful, pasture and mobility were limited since the large irrigated area had been laid out across traditional migratory routes and crops were now grown throughout the year. Many trees, which once provided shade and fodder in the dry season, were removed in the project as were valuable henna hedges.

Pastoralists faced more acute problems downstream and east of Kano where, instead of the annual flood, the river regime is now controlled by the Tiga Dam. The river channel has deepened and narrowed and *fadama* land, dependent on the annual flood for water and silt, is now left high and dry above the channel. Fulani in this area complained bitterly about the effects of the Tiga Dam: 'the dam has denied us water in the fadamas and has compelled us to reduce the size of our herds to a size manageable by well water'. Once perennial streams are now, in the words of one herdsman, 'choked by sand'. Many pastoralists, whilst expressing pride in their herds and lifestyle, gave the impression that they were gradually succumbing to the pressures to settle and adopt a mixed farming economy.

The patterns and processes associated with the general intensification of agriculture around Kano, together with the effects of government intervention in the form of the Tiga Dam and the Kano River Project, have severely constrained the lifestyles of

Case study G (*continued*)

Fulani pastoralists. With the additional impact of drought, it is likely that pastoralists have suffered more than any other group, since their long-established ability to cope with shortages of water and pasture, through migration, has been progressively restricted. The Fulani way of life and its changing character must be appreciated and provided for in future development plans, to ensure that living standards do not decline further. However, it is likely that the increasing sedentarization of pastoralists and their herds is actually what the government would like to achieve.

Case study H

Villagization in Tanzania

When Tanzania gained its independence from Britain in 1961, well over 90 per cent of the country's population lived in scattered rural settlements. Agriculture's contribution to the nation's gross domestic product was falling and most farmers were engaged in growing food for family subsistence using traditional low-level technology. Inequalities were increasing in the distribution of income and the provision of basic services such as health care and education. The country was also becoming increasingly dependent on foreign loans and grants.

Less than six months after the end of colonial rule, in 1962, Julius K. Nyerere, President and leader of the ruling party, TANU (Tanganyika African National Union) published a pamphlet, *Ujamaa – The Basis of African Socialism*, in which he urged a return to 'traditional values', according to which everyone had a right to be respected, had an obligation to work, a duty to assure the welfare of the whole community and the common ownership of basic goods. Later in 1962, in his inaugural address as President, Nyerere introduced the idea of villagization, emphasizing the importance of developing the agricultural sector, which, he argued, could only be achieved if farmers lived in villages. He said,

Case study H (*continued*)

> Unless we do . . . we shall not be able to use tractors; we shall not be able to provide schools for our children; we shall not be able to build hospitals, or have clean drinking water, it will be quite impossible to start small village industries, and instead we shall have to go on depending on the towns for all our requirements; and even if we had a plentiful supply of electric power we should never be able to connect it up to each isolated homestead.

Five years later, the Arusha Declaration of February 1967 was a landmark in Tanzanian, and indeed African, political history. Nyerere announced the nationalization of banks, trading organizations and the largest multi-national corporations operating in the country. He also called for a halt to the accumulation of private wealth by party and government leaders. In essence, the Arusha Declaration outlined the key elements of socialist development, stressing the use of local ideas and resources. This was reinforced in September 1967 with Nyerere's paper on 'Socialism and Rural Development', whose Swahili title was *Ujamaa Vijijini*, which literally translated means 'socialism in the villages'. In this paper Nyerere again rejected rural capitalism and turned the *ujamaa* (familyhood) of his 1962 paper into a national policy such that rural workers had the responsibility to establish and/or encourage *ujamaa* villages, building on traditional mutual aid in the extended family of Tanzanian society.

The main aims of the new rural development strategy were:

1 The establishment of self-governing village communities to improve living standards by providing social infrastructure and producing and marketing crops and livestock.
2 Better use of rural labour, taking advantage of economies of scale to increase communal production, though the emphasis on communal production was later reduced.
3 The dissemination of new values and avoidance of exploitation. Village councils would oversee land reform, allocating land among private cultivators.
4 Mobilization of people for national defence by using the villages as para-military organizations.

Case study H (*continued*)

Help was to be available from government and other organizations to explain the underlying principles of *ujamaa* villages, to promote good leadership among farmers and commitment among government leaders and officials and to aid in the planning of village sites, food cultivation and service provision. Village development was encouraged through a series of 'operations', but campaigns by regional party and government administrations were often carried out hurriedly, with insufficient planning, little consultation with the people and a limited understanding of existing farming and pastoral systems. The President himself took a personal interest in the formation of *ujamaa* villages, for example, by initiating 'Operation Dodoma', a government-planned programme to move all people in that Region into villages. Consequently, the number of villages in Dodoma increased from 75 in 1970 to 246 in 1971.

On 6 November 1973 Nyerere announced that all rural Tanzanians would have to live in villages by the end of 1976, and at the same time some fundamental policy changes were made. Emphasis on communal production, for example, was dropped in favour of block farming, designed to promote economies of scale and village production planning, but not requiring a whole-hearted commitment to the principles of *ujamaa*. From this time the new villages were called 'development villages', and could serve as multi-purpose cooperatives, taking on all crop marketing and credit functions. This major thrust of compulsory villagization between 1973 and 1976 was probably the largest resettlement effort in the history of Africa. Although most people did not have to move over long distances, they often had to agree to abandon their residence and land. People were instructed to move to the nearest existing village or trading centre and political efforts were made to create *ujamaa* villages of 250 or more families out of the enlarged units. Initially, people were moved into new villages as a result of persuasion, and the prospect of services such as water, schools and dispensaries often served as an incentive. However, sometimes promised services, including food supply, were inadequate. Villages were at first concentrated in the poorer regions of

Case study H (*continued*)

Tanzania, and by September 1974 there were 5000 villages with 2.5 million members, representing 20 per cent of the country's total population.

In 1974 the government decided to abandon its strategy of voluntary inducements in favour of compulsion. Between May and December 1974, Operation Sogeza (moving) implemented this decision, so that by 1975 a large majority of the rural population was resident in villages. From 1975 a series of bad harvests and food shortages encouraged the government to increase agricultural production and peasants were required to plant a minimum of three acres of food crops and one acre of cash crops. By the late 1970s, however, many village-based intiatives had collapsed and the future looked unpromising.

The main problems encountered by the villagization programme were:

1 For most people in most villages private farms still remained the primary focus of interest and there was some conflict as to how much labour and attention should be given to communal activities. The first five years of *ujamaa* did not have any significant impact on peasant agricultural production, nor did the productivity of the land improve. Yields per hectare in communal farming were well below those of private farming, and there was often confusion over the distribution of income from the communal farm. Communal work actually ceased in some villages by the early 1970s. The bigger the village, the more difficult it was for members to achieve a common sense of purpose. Petty-capitalist farmers often did best, being more willing to take risks and modernize their farming. Poorer peasants often chose these more successful farmers as their village leaders.

2 In many *ujamaa* villages the absence of a reliable system of financial control caused problems, perhaps due to the low level of education of leaders and villagers. Embezzlement of funds was common and administrative and technical staff often failed to serve the villages loyally and intelligently.

3 Impressive progress was achieved in the provision of schools, dispensaries, water supplies and other rural infrastructure, and

Case study H (*continued*)

marked improvements in health care and adult literacy were recorded. However, schools and dispensaries often lacked the necessary basic supplies. Also, in spite of better rural service provision, young people were still attracted to education and work in the towns, which was often supported by parents because it could yield additional family income.

4 The principles of *ujamaa* and communalization frequently conflicted with well-established networks of social institutions such as women's groups and cattle-owner associations. Furthermore, the social effects of villagization must have been traumatic at times, particularly where force was used.

5 There is still much debate about whether villagization actually reduced inequality within and between villages and households.

6 Where tractors and other machinery were introduced, there were problems of maintenance and lack of fuel, particularly after the 1974 oil crisis.

7 The environmental knowledge that peasants had prior to villagization was often invalid in the new settlements where soil conditions and other factors of production were different. There was concern in some areas that nucleation could lead to erosion as the carrying capacity of the land was exceeded.

8 Cooperatives were completely abolished in 1976, after which the buying of crops and provision of farm inputs became very unreliable.

For the majority of peasants, the decade after the Arusha Declaration was not so much a period during which *ujamaa* had failed, but a period when they had been subjected to numerous government directives and orders without witnessing much economic development. The villagization programme generated massive resentment amongst peasant farmers. Most writers agree that the villagization programme was a failure and indeed the Tanzanian Government has since reversed many of its earlier policies.

Key ideas

1 Africa is still mainly a rural contintent. Agricultural production and pastoralism employ the majority of Africa's people and their ᵣutput makes a significant contribution to domestic production and export revenue.

2 Food production per capita has been declining in Africa in recent years, whereas in Asia there has been a steady upward trend in food production.

3 Although feeding the household is a key objective in traditional African farming systems, many farmers also produce marketable surpluses.

4 African food production systems are closely geared to the different seasons and their main input is human labour, much of it by women.

5 Pastoralism may be the best adaptation to environment in marginal-arid and semi-arid areas.

6 Rural non-farm employment can make an important contribution to the incomes of rural households.

7 Many large-scale development schemes have had a significant negative impact on environment and society and their appropriateness must therefore be questioned.

5
Urban Africa

Urbanization in tropical Africa

The growth of towns in tropical Africa has been recent and rapid, yet the bulk of the population still lives in rural areas. By the year 2000, Africa will have the distinction of being the only continent with a majority rural population. But population growth rates in tropical Africa now exceed those in most other parts of the world and the urbanization of its population, whilst lagging behind trends in Latin America and Asia, is now proceeding more rapidly than anywhere else. While the population of Africa as a whole is expected to become only 42 per cent urbanized by the year 2000, its 1985 urban population will double by that time, with an average annual growth rate of 4.6 per cent compared to 3.7 per cent for other less developed regions and 1.1 per cent for the more developed regions (see Figure 5.1). This trend is worrying in such a poor region with limited resources for providing the necessary urban services.

The fact that this urban growth is recent is shown by an examination of the situation in 1960, on the eve of independence for most African countries. In that year tropical Africa only had three cities with over 500,000 inhabitants; Ibadan and Lagos in Nigeria and Kinshasa in Zaire. Some fourteen countries had no city with 100,000 inhabitants. Twenty years later, in 1980, there were no fewer than twenty-eight cities in tropical Africa with 500,000 or more people. The main factors causing this rapid urban growth have been high rates of natural increase of both

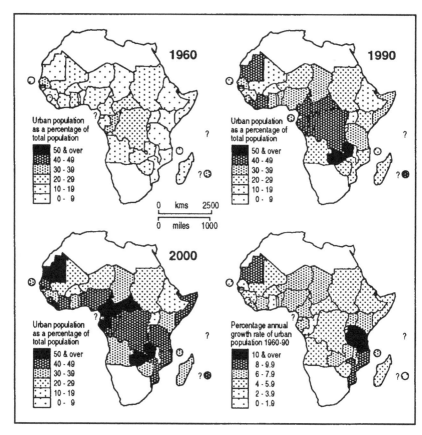

Figure 5.1 Urban population as a percentage of total population

urban and rural populations, the incorporation of surrounding villages into built-up areas and growing and more permanent rural–urban migration. The latter has had a particularly significant impact on urban growth. For many people in poor rural areas lacking basic facilities and receiving low prices for their crop production, the towns, and particularly the large cities, are seen as havens of opportunity where jobs and wage employment can be obtained. It has been estimated, for example, that in the ten-year period between 1953 and 1963, 644,000 people migrated to Lagos in Nigeria, accounting for 75 per cent of the city's total population growth.

Urban growth has been most rapid in the capital cities and other large

Plate 5.1 New bank in Bamako, capital of Mali.

cities. Of the tropical African countries with a coastline, the vast majority have coastal capitals, some notable exceptions being Kenya (Nairobi) and Zaire (Kinshasa). Until recently all the coastal countries of West Africa had coastal capitals, though the pattern has been distorted recently with the creation of new capitals at Abuja in Nigeria and Yamoussoukro in Ivory Coast. Most capitals of tropical Africa are also primate cities with their populations more than twice the size of the second city. In many countries the urban system is dominated by one or two cities. For example, in 1980 it was estimated that 83 per cent of Mozambique's urban population lived in the capital, Maputo, and 80 per cent of Guinea's urban population lived in Conakry, also the capital city. Nigeria provides a good and rare example of a country with a well-developed urban system, with only 17 per cent of the urban population in 1980 living in Lagos, the capital and largest city. With the detailed results of the 1991 Nigerian census still awaited, projections based on the last census in 1963 suggest that by the mid-1980s there were no fewer than six cities with more than a million people and fourteen with more than 500,000.

Phases of urban growth

Although tropical Africa has experienced its most rapid period of urban growth since 1960, it would be wrong to neglect the presence of towns in the pre-colonial period. The most important areas of pre-colonial urban development were in the Ethiopian highlands, along the coast of East Africa and in two major zones of West Africa, in the forest belt and further north in the savannah region. The fact that West Africa was the most urbanized sub-region of tropical Africa in 1960 testifies to the significance of some large indigenous towns and cities. Before the colonial intrusion into West Africa, there existed a series of successively important empires and kingdoms such as Ghana, Mali, Songhai, Oyo and Benin, based on centralized political and economic power located in cities. In the savannah region south of the Sahara desert, a line of important towns developed at the southern end of a flourishing trans-Saharan trade. Initially, salt was brought southwards from the Sahara and exchanged with gold, but later the trade diversified with European and Arab goods being exchanged for commodities such as slaves, skins, gum and spices. The town of Djenné was founded as early as the ninth century, whilst Gao, Kilwa and Kano all date from the late-tenth century. Others followed, such as Zaria (1095) and Timbuctu (1100). Timbuctu became an important religious and educational centre with the first university in West Africa. Further east, in what is now northern Nigeria, Kano had 75,000 inhabitants or more in the sixteenth century. However, the coming of the Europeans and the development of the Atlantic slave trade in the eighteenth century switched the trading emphasis in West Africa towards the coast, and the trans-Saharan trade and many associated towns declined.

Other states with important settlements existed in the forest region of West Africa. The Ashanti kingdom, founded on gold mining in what is now south-western Ghana, had its seat of government at Kumasi and exercised considerable power until it was conquered by the British in 1873–4. Another forest-based state was Benin in southern Nigeria, famed for its bronze artwork, which flourished in the sixteenth and seventeenth centuries. Perhaps the most famous indigenous towns in West Africa are those of Yorubaland in southwestern Nigeria. These walled towns, such as the Yoruba spiritual capital of Ile-Ife established by the tenth century and the political capital Old Oyo, drew their wealth from productive farming hinterlands and a wide variety of trades and crafts. Wars between Yoruba towns in the early-nineteenth century

Plate 5.2 Main post office in Bamako, Mali. A fine example of French colonial architecture.

resulted in the creation of new defensive towns such as Ibadan in 1830, which by 1865 was the largest Yoruba city and later became the largest city in tropical Africa until the 1960s.

Most of tropical Africa's urban centres are of colonial origin dating from the late-nineteenth and early-twentieth centuries. These towns, though often initially centres of trade, soon became centres of administration and political control. Many were located on the coast, along important lines of communication or near mining centres. Accra, now the capital of Ghana, was initially a number of coastal slave forts, but grew rapidly from 1877 when the British moved their administration from Cape Coast. Nairobi in Kenya was established almost accidentally as a railway camp in the 1890s in an area lacking in large settlements. The mining towns of Zambia were mostly developed in the late 1920s when rising copper prices made it worthwhile to invest in mines at Chingola, Kitwe, Luanshya, Mufulira and Ndola.

Many colonial towns show a clear distinction in their structure between functional and residential zones. Within the residential areas

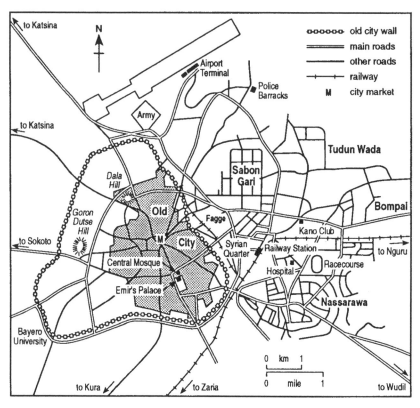

Figure 5.2 Kano, Nigeria

there were further divisions according to race and class. French towns typically had tree-lined boulevards, parks and pavement cafés. Dakar, capital of Senegal, grew rapidly in the colonial period, acquiring many features of a typical French city and assuming the position of capital of French West Africa. In areas of British colonization, the racecourse, golf course and club were key features. This is clearly seen in the city of Kano, northern Nigeria, where, with the advent of colonialism in 1903, a European residential area was established to the east of the old walled city with wide roads, spacious housing plots and recreational facilities (see Figure 5.2). A quite separate area for non-northern Nigerians, known as Sabon Gari, was laid out on a grid-iron pattern with buildings at a much greater density. A further sector, west of the railway station and close to the developing central business district, was settled by

Syrian and Lebanese traders. Thus, within Kano there is a series of clear structural divisions, based on the ethnic and social characteristics of their original inhabitants. In other cities too, quite separate settlements built by and for the colonial masters can be seen. For example in Freetown, Sierra Leone, Hill Station was built in a location with a fresh and comparatively tolerable climate and linked to Freetown below by railway. Some colonial cities lacked any indigenous element, and were established primarily as places for Europeans to live. Nairobi, Harare, Lusaka and Bulawayo were built on European lines and were established primarily to provide services to European settlers in the surrounding areas.

Since independence, urban growth has quickened and some towns have outgrown their sites, such as Lagos, which is permanently congested and has little space for new development. In the case of Nigeria there has also been a desire for a more centrally located capital. In 1976, the Nigerian government announced its intention to transfer the capital to Abuja in the middle-belt, a relatively underpopulated region. By the end of 1993 it is anticipated that the movement of government ministries to the new capital will be complete (see Case study I). Also in West Africa, Ivory Coast is developing a new national capital at President Houphouet-Boigny's village of Yamoussoukro. In Malawi, the decision to relocate the capital in Lilongwe, close to President Banda's birthplace, was taken in 1968, but it was not until 1975 that it actually succeeded Zomba as capital. It was argued that a more central location in an area of good agricultural potential was needed, with the possibility of locating a new international airport nearby. South Africa provided much financial and technical assistance for the project, as well as drawing up a master plan. By 1989, however, there had been little development in the new capital apart from government buildings and the airport. Tanzania has also decided to move its capital from Dar-es-Salaam on the coast to a location 450 km westwards at Dodoma, in the hope of stimulating economic and agricultural development in the country's underdeveloped heartland. However, rising building costs have meant that progress on building the new capital has been slow, and it is difficult to predict when the transfer of responsibilities will occur.

Rural–urban links and peri-urban pressures

Rural areas surrounding the rapidly growing towns of tropical Africa are gradually being incorporated into the urban system and their character is changing. These peri-urban areas are often zones of intensive market-orientated food production where rural producers and urban consumers are closely juxtaposed. They are areas where once-rural villages have lost their land to urban development and existing farmland must now be used more intensively to supply profitable urban markets. Where farmland can still be obtained in the peri-urban fringe, urban dwellers may undertake part-time farming to supplement their food supply and income. There is evidence from some towns that capitalist farmers, relying mainly on hired labour, are buying land in peri-urban areas, usually along major roads, to supply urban food markets. Commuting from the countryside to the town in search of paid employment is also common, particularly where dry season cultivation is not possible. These peri-urban areas show a wide variety of production and exchange relations with considerable linkage between urban and rural areas expressed in two-way flows of goods, labour and money.

Studies of a number of tropical African cities have revealed intensively cultivated peripheral areas developing in the post-war period, and particularly since independence. Urban dwellers, particularly women, grow crops to supplement household food supply, whilst men often concentrate on cash crop production. For example, women from the Makelekele quarter of Brazzaville, Congo, some of them coming from affluent households, grow cassava, whilst poorer or unemployed men grow salad crops to sell to better-off households, hotels and restaurants. Dar es Salaam, with a population of over 1.3 million, provides a major commercial opportunity for peri-urban farmers to produce food crops for consumption within the city. The peri-urban zone starts at about 20 km from the city centre and in the early 1980s this agricultural expansion was adopted mainly as a survival strategy when urban incomes and living standards were squeezed. However, since the mid-1980s the continued expansion has been geared more towards commercial sales rather than satisfying immediate household needs. Over 93 per cent of farmers grow cassava, whilst other popular crops include bananas, maize, pawpaw and a variety of fruits and vegetables. The average size of farms is 1.6 ha, and intercropping is widely used. Since the 1960s a large number of people have actually settled in the peri-urban zone, and in recent years land has acquired more of an investment status, putting pressure on smaller farmers.

The mountain villages close to Freetown in Sierra Leone and horticultural areas near Banjul and tourist hotels in The Gambia are further examples of intensive cultivation of fruit and vegetables in peri-urban areas. In Kano, Nigeria, three sets of peri-urban pressures can be identified. First, there is competition for land between rural and urban dwellers, particularly over *fadama* land, where the water table is near the surface and irrigated vegetable production for the urban market can continue through the long dry season. Second, there is competition for markets for foodstuffs, but also for firewood and manure, the latter for fertilizing the intensively-farmed fields. Third, competition also exists for non-agricultural employment such as home-based craft industries, firewood sales or work in Kano city.

Too often in the past rural and urban have been seen as quite separate, but in reality there are many complex interrelationships between them. Movements of people, income and capital, goods, social transactions and the transmission of ideas, innovations, power and authority are all evident to a greater or lesser extent. As centres of power and wealth, the towns and cities of tropical Africa might be seen as parasitic, dependent on the rural hinterland for food, labour and other commodities such as firewood. It is a fact that African farmers receive poor prices for their food crops. The principal beneficiaries are not the producers, but are either the trading middlemen or the urban consumers. Many states control the prices of basic foodstuffs in towns, since there have been numerous instances of urban rioting and political instability following price rises. It is important that in formulating future development plans urban–rural linkages are fully appreciated and neither urban nor rural development should be treated in isolation.

The housing problem

The rapid growth of towns and cities in tropical Africa has put a tremendous strain on housing and service provision. Hard-pressed urban authorities are faced both with streams of newcomers from the rural areas and existing urban populations which are growing rapidly. Most Africans, even those who have lived for a long time in towns, retain strong links with their 'villages' and periods of urban living are often viewed as a necessity and an opportunity to earn money which can then be invested in a house, the extended family or the community back home. For many, life in the town is seen as a temporary phase in their lives.

There is a real housing problem in many large African cities where people are forced to live in cramped and unhealthy conditions. A survey of Lagos, Nigeria in 1976 revealed an average occupancy rate of 4.1

Plate 5.3 Unplanned housing close to the centre of Freetown, Sierra Leone. The stream in the picture is heavily polluted with sewage. Water from the stream is used for irrigating vegetable crops which are then taken to one of the city's thriving markets.

persons per room. Private rented accommodation, built on plots leased from a municipal or traditional authority or occupied illegally, is most common in tropical Africa. Frequently, the owner lives in the house, renting out rooms to others who may or may not be part of his or her own extended family. For most households rent forms a large part of household expenditure. Other types of ownership by private owners or governments and institutions are much less common.

Residential density in African cities is generally not as great as in some Asian or Latin American cities due to the existence of spacious, often colonially-planned suburbs and the unpopularity of high-rise housing. However, at least a third of Africa's urban population live in what might be termed 'slum' or 'squatter' settlements, and these are growing at the rate of 15 per cent per annum, causing them to double in size in six years. The term 'slum' has been defined by the United Nations as, 'areas of authorised, usually older, housing which are deteriorating, in the sense of being underserviced, overcrowded and dilapidated'.

Squatter or 'spontaneous' settlements have many of the same characteristics as slums, but with the additional disadvantage that the land is
being used illegally. Whereas slums tend to be central, squatter settlements are usually peripheral, since squatting in central areas would soon
be stopped by the owners, except on marginal land of no immediate use.
New migrants tend to move to more central slums near to possible job
opportunities in the city centre. More established residents, however,
who have made a decision to stay, may prefer to squat on the edge of
town in their own family houses. In many African cities some central
slums are occupied by long-standing urban residents who continue to
live in their family houses which may have been handed down through
several generations. It is not uncommon for particular areas of African
cities to be dominated by single family or tribal groups who may have
been urban residents for some time and act as hosts to members of their
group who may be visiting from other towns or the home village.

Housing policy alternatives

Governments and urban authorities adopt different attitudes to slums
and squatter areas. Sometimes it may just be too expensive to do
anything about them, since the cost of clearing poor housing and
rehousing the inhabitants can be prohibitively high. Even if authority-
owned housing can be built, tenants may find it difficult to pay what are
regarded as economic rents to bring an adequate return on the investment. In other situations, governments are more hostile to areas of poor
housing, seeing them as symbols of their lack of success in providing for
the urban population and modernizing the built environment.
Politicians and wealthy urban elites may consider these areas unsightly
and press for their demolition. Alternatively, there have been cases
where urban authorities have provided help to slum and squatter
dwellers in upgrading their housing.

One such example of upgrading is seen at Chawama, a squatter
settlement in Lusaka, the capital of Zambia, which has developed since
1952. By 1974 there were seventeen squatter settlements in the city,
accommodating about 40 per cent of Lusaka's population. Chawama
was the second largest settlement, with 25,000 inhabitants. In the early
years the government took a negative view of squatter settlements, but
more recently they have gradually been recognized with the introduction of certain basic services. About a quarter of families in 1974 had
lived in Chawama for over eight years and three-quarters of the working

men and women were employed mainly in construction and other industries. Household income has risen substantially since independence. A key move in the long-term development of Chawama was when in 1966 the government bought the land from the white South African landowner. In the same year the wells in many Lusaka squatter settlements dried up and, after protests by the women, a piped water system was installed, though it soon proved inadequate for the needs of the growing settlement. A surfaced road was built by the residents in a community self-help project, where tools and materials were supplied by the city authorities. Afterwards, in 1969, the city council introduced a bus service to the city centre. A market was also built and jointly managed by community and government departments. The only government primary school in a squatter settlement was opened in Chawama in 1969, and in the same year a pre-school group was established. After much protest, improvements in Chawama were gradually achieved by the residents, but so much depended on them putting pressure on the authorities at the right times, usually when it seemed politically expedient for the government to help.

What can be done about the housing situation in tropical Africa's growing cities? The modest achievements at Chawama are part of a massive programme of squatter upgrading in Lusaka, involving first the granting of legal occupancy and then the installation or improvement of basic services. Such projects have had some success, since with the legalizing of squatter settlements the communities themselves are more willing to get involved in long-term improvement.

As we have already seen, the provision of government housing for low-income groups has been fairly limited in tropical Africa, and although some provision has been made in cities such as Harare and Lusaka, it is often a long distance from places of work. The building of council estates has also been undertaken in Nairobi since 1945, such that by 1963 almost half the city's population lived in government housing. However, the problems of high building costs and rents beyond the means of the urban poor, have encouraged governments to concentrate on either slum and squatter upgrading or site and service schemes.

Site and service schemes are usually located on the outskirts of cities, where urban authorities clear and lay out areas with housing plots, roads, piped water, sewerage and electricity. Households and communities are then encouraged to build their own homes on the plots, often with government assistance in obtaining and purchasing building materials. Such projects often represent a continuation of similar colonial housing policies. Site and service schemes are found in many countries in tropical Africa, including Kenya, Ivory Coast, Tanzania, Zambia, Malawi, Senegal and Sudan and have received encouragement and support from the World Bank. The first scheme supported by the World Bank was in 1972 in Pikine on the outskirts of Dakar, Senegal. The Bank has also been involved in a number of schemes in Lusaka which have provided over 11,500 new housing units. One common problem of site and service schemes is that with their peripheral location, a long and difficult journey to work is often necessary unless employment opportunities can be located near the new settlements.

The urban environment

The speed and unplanned nature of urban growth in tropical Africa has generated many environmental problems, the most serious of which are a lack of piped water systems for homes and businesses, inadequate sanitation and sewerage provision and considerable water and air pollution. Government authorities and aid agencies need to give more serious attention to improving the environmental situation in urban areas. A survey of 660 households in Dar es Salaam in 1986–7 revealed that 47 per cent had no piped water supply either inside or immediately outside their homes, while 32 per cent had a shared piped water supply. Of the households without piped water, 67 per cent bought water from neighbours, while 26 per cent drew water from public water kiosks or standpipes. Sewage disposal was totally inadequate, with only 13 per cent of dirty water and sewage being regularly removed. In Khartoum, Sudan, the municipal sewerage system serves only about 5 per cent of the urban area and the system regularly breaks down, leading to waste discharge into the river Nile or on to open land. Kinshasa, Zaire has no sewerage system and about half the population (1.5 million) have no piped water supply. The collection of household waste is only undertaken in a few residential areas of the city. Elsewhere, waste is put out on the road, on illegal dumps or in storm-water drains.

The provision of basic services by urban authorities is expensive, difficult and disruptive. Possibly the best alternative is for governments and local community organizations to work together in low-cost projects such as draining stagnant pools, installing pipes, drains and access roads. There is also a need to promote educational programmes on health prevention and personal hygiene.

Urban employment: the informal sector

The greatest proportion of rural–urban migrants come to the city in search of employment, but in many countries the pace of urban growth is outstripping the capacity of the economy to provide jobs. A major problem in investigating urban employment patterns in tropical Africa is the lack of data coupled with the problem of defining terms such as 'employed' and 'unemployed'. Even in certain high-income countries, there is an ongoing debate about the accuracy of unemployment statistics. But in many African countries there is no detailed register of unemployed persons and state unemployment benefits are rare. Those who appear to be unemployed may be more accurately termed 'under-employed', since there are many opportunities for casual paid work in African cities.

Major distinctions are often made between formal and informal sectors of the labour force and between large- and small-scale, wage- and self-employed. Labour markets are much less segregated than in industrialized countries and it is common for people to have more than one job and to cross the boundaries between different sectors, sometimes on a regular basis. Construction workers, for example, may alternate between working for themselves and for large firms. Trading in different commodities at various levels is undertaken by a wide range of people, sometimes by those who are already in full employment as a source of additional remuneration.

The term 'informal sector' was popularized by research in Ghana and the International Labour Organization report on Kenya in the early 1970s. Since then many people have attempted to define and classify the informal sector and there have been numerous case studies. It was initially argued that informal sector employment is relatively easy to enter with no formal qualifications, it is small-scale and often family-based, makes use of adapted local technology and resources and is labour intensive. However, the formal/informal dichotomy has been criticized for being too simplistic and rigid. Some have suggested that even within the informal sector there are many sub-sectors. There is little in common, for example, between the legitimate business of a tailor or a metalworker, a street hawker or a child who guards parked cars, and the illegitimate work of a prostitute, a drug-pusher or a pickpocket. The simple dichotomy further obscures complex linkages between formal and informal sectors and the ways in which particular enterprises change over time, perhaps moving from being illegitimate to legitimate or from being informal to being formal. If nothing else, the informal/formal sector dichotomy has generated much debate and provided a framework for empirical research. Some recent studies have revealed positive aspects of informal sector enterprise and innovation which might play an important role in future personal, community and national development.

In Nairobi, the government's negative attitude to slum areas in the 1960s and 1970s has changed to one in favour of improvement. Mathare Valley, with a population approaching 200,000, has the largest concentration of informal enterprise in Nairobi, with some 20 per cent of enterprises being over ten years old. Many were started with less than $US50 obtained from savings, sale of assets or loans from friends and relatives. Eighty per cent of enterprises surveyed in 1989 were in trade and commerce, 11 per cent in manufacturing and 9 per cent in services. Virtually all the businesses were sole ownerships with little scope for employing extra labour. Eighty-six per cent of operators were under forty-nine years old and about half were women, mainly selling green-groceries. Business operators generally had seven or less years of formal education and few kept written records. Enterprises mainly sold products directly to customers from the local area, but some had clients outside Mathare Valley. Average monthly profits ranged from $155 for provision kiosks and butchers to $86 for other types of retailing. These profits compared favourably with the legal minimum monthly wage in Nairobi of $28 and average monthly wages of $98 in Kenya's private sector and $110 in the public sector. Furthermore, in addition to profits from these enterprises, almost a third of operators earned additional income from such sources as part-time wage employment provided by other operators, spouses and other members of the family, from other businesses, especially the illegal brewing and sale of alcohol, from farming for those with land, from rents for landlords and from friends and relatives. This additional income averaged $80 per month which, when added to the enterprise profits, was adequate for household needs and usually left some for savings. There was considerable interaction between the informal enterprises in Mathare Valley and other enterprises outside the area. Tailors, for example, went outside the area to buy thread, cloth and buttons. Infrastructural facilities were minimal, with poor roads, under 10 per cent of enterprises connected to electricity, 4.5 per cent connected to piped water and only 2.5 per cent connected to a sewerage system. Although the majority of enterprises operated from a fixed location, about 25 per cent were located in the open, while others were based in temporary corrugated iron sheds, timber huts or were operated from the house. Operators complained about the lack of infrastructure, the problem of obtaining credit, high rents and insecurity of tenure because many business premises were temporary and rented. Only 11.5 per cent of operators partially or fully owned their business premises and were worried they might be moved by the authorities because of fire risk or other reasons.

Many of the features of Mathare Valley are typical of informal enterprises elsewhere in tropical Africa. A wide variety of dynamic and innovative activities frequently exist which can play a role in education and skills transmission, in generating income and employment and in providing a range of products and services. Governments and urban authorities need to consider carefully how such activities can be further supported and stimulated to enhance the role of the informal sector in generating national wealth.

Formal sector employment

It would be incorrect to assume that formal sector enterprises are necessarily larger than those of the informal sector, but this is often the case. Formal sector enterprises usually employ more people, are characterized by a greater degree of permanence, often have a substantial workplace, are more likely to be recognized and enumerated by the urban authorities, and generally have wider links throughout the country and internationally. But the variety of enterprises and employment opportunities in the formal sector is as great as in the informal sector.

An employment survey undertaken in Kenya in 1979 revealed that about 20 per cent of Nairobi's labour force was employed in manufacturing, 12 per cent in construction, 30 per cent in commerce and finance, 8 per cent in transport, 16 per cent in administration and social services and 10 per cent in domestic service. This pattern is not untypical of other capital cities (see Case study J). The importance of administrative and service employment, including domestic service, is a common feature of many tropical African cities. Administration, commerce and finance are particularly important in most capital cities, notably Yaounde in Cameroon, Nouakchott in Mauritania and Kigali in Rwanda. Compared with certain regions of Europe, there are relatively few industrial towns. The mining towns of the copperbelt of Zaire and Zambia are exceptions. In Zambia, towns such as Chingola and

Mufulira owe much of their existence to copper mining. In Sierra Leone, the rapid growth of Koidu has occurred since diamonds were discovered there in the 1930s. Elsewhere in tropical Africa, however, mining employs relatively few people.

Little modern manufacturing industry was established in tropical Africa during the colonial period. The European powers viewed their colonies primarily as suppliers of raw materials and cheap labour and industrial development was generally discouraged. However, traditional craft-based industries pre-date the colonial period and survive in many countries to this day. Unfortunately, they have suffered from external competition and were often neglected by colonial and early independent governments. But recently some African governments have attempted to stimulate these traditional craft-based industries, making such items as cloth, wooden carvings and metalwork. These products are primarily for domestic consumption and also for the growing tourist industries (see Case study K). The craft industries of the Yoruba towns in south-western Nigeria are well documented. The family compound is the main unit of production and emphasis is on use of family labour and local materials. A study of the town of Iseyin in the 1960s showed that 27 per cent of the male population were employed in the weaving industry and numerous other people worked in the related carpentry industry making looms, shuttles and bobbins for the weavers. Also in Nigeria at Ilorin, there is a thriving pottery industry and a narrow-loom weaving industry catering mainly for the luxury market. There are many other examples of traditional craft industries both in towns and rural areas throughout tropical Africa.

Industrialization was seen as a key priority at the time of independence. African countries were keen to emulate their former colonial masters and saw an industrial revolution as a key to their future development. However, tropical Africa's industrial performance since the early 1960s has been very disappointing. This performance can be measured by Manufacturing Value Added (MVA), which is the difference between the gross output of the manufacturing sector and the sum of physical purchases and service inputs before any provision is made for depreciation. The whole of Africa's share of world MVA rose fractionally from 0.7 to 0.8 per cent during the 1960s, but has since fallen to 0.3 per cent. In 1989 for the countries south of the Sahara (i.e. including the relatively more industrialized South Africa), manufacturing contributed only 10 per cent to the region's Gross Domestic Product (GDP), compared with over 30 per cent for industrial market economies and 24 per cent for other developing countries. In tropical Africa in 1989, five countries accounted for almost 60 per cent of sub-Saharan MVA. These were in order of importance, Nigeria, Zimbabwe, Kenya, Ivory Coast and Ethiopia. Whilst the last four of these countries had a small positive growth in MVA between 1980 and 1989, ranging from 0.3 per cent in Ivory Coast to 3.5 per cent in Kenya and Zimbabwe, Nigeria actually experienced a negative –5.0 per cent growth rate. Factors which have prevented the further development of manufacturing industry include the low per capita income level and the size of population, which affects the size and purchasing power of the internal market, the absence of supportive infrastructures in the form of transport and reliable power supplies, the nature of the existing industrial base and the countries' poor resource endowment. In addition, African countries have suffered from a heavy debt burden, adverse terms of trade and the poor performance of their agricultural sectors, which has affected consumer demand, the level of farm inputs and supplies of raw materials.

Tropical Africa's manufacturing industry is dominated by two subsectors; food processing (including beverages and tobacco) and textiles. These account for more than half MVA in the majority of countries, except Nigeria and Zimbabwe. From the early post-independence period, industrial development has emphasized import-substitution, but since much of this industry depended heavily on imported inputs, problems were encountered after 1974 with scarce foreign exchange, falling commodity prices and rising oil prices. Import substitution failed to make full use of domestic resources and de-industrialization actually occurred in Benin, Congo, Ghana, Madagascar, Tanzania, Togo and Zaire, as well as Angola and Mozambique, where industrial failure was largely associated with the revolution in Portugal in 1974 and the belated granting of independence in 1975. Manufacturing has also failed to generate jobs. It is estimated that between the mid-1960s and late-1980s only 950,000 jobs were created in manufacturing in thirty-two countries and 80 per cent of the new jobs came in a handful of countries, notably Nigeria, Kenya and Zimbabwe. However, significant job losses occurred in the 1980s in Mozambique (60,000), Zaire (16,000), Nigeria (140,000), Tanzania (12,000) and Zambia (7,000).

The future for industry in tropical Africa looks very uncertain and will depend to a large extent on progress in the agricultural sector. If this occurs then it could lead to an upsurge in the processing of agricultural products for both domestic and overseas markets and in the manufacture of basic consumer goods and inputs for the agricultural sector.

Case study I

Urban growth in Nigeria

Nigeria has been called 'the giant of Africa'. It has a larger population than any other country in tropical Africa, has the largest economy and in recent years has played a leading role in African politics and diplomacy. It also has some of the largest towns in tropical Africa and is experiencing a range of pressures associated with rapid urbanization.

Provisional figures from the 1991 census indicate a total Nigerian population of 88.5 million, rather surprisingly some 30 million fewer than the World Bank estimate, but nevertheless a substantial increase from the 1963 figure of 55.7 million. The average annual population growth rate of 3.2 per cent between 1980 and 1990 is projected to fall slightly to 2.8 per cent in the period up to the end of the century. The World Bank estimates Nigeria's urban population to be 35 per cent of the total population, producing a figure based on the 1991 census of 31 million. This probably means there are more urban dwellers in Nigeria than in all the other West African towns and cities put together. In 1980 Nigeria had no fewer than 26 cities with a population greater than 100,000, significantly more than any other tropical African country. The proportion of the total population living in cities of over 20,000 people was estimated to be 4.8 per cent in 1921, but this figure had risen to 19.2 per cent by 1963. It has been estimated that in recent years urban population has been growing at an annual rate above 6 per cent. Lagos, with a current population over 3 million, overtook Ibadan as the largest city in Nigeria between 1953 and 1963 and has experienced phenomenal growth, further enhanced during the civil war (1967–70) when it was outside the zone of hostilities and oil companies moved their headquarters there from Port Harcourt. Lagos may now well be the largest city in tropical Africa.

The distribution of Nigeria's largest cities is rather uneven (see Figure 5.5). The south-west is the most urbanized region, a legacy of the long tradition of urbanism which has been a feature of Yoruba culture dating back to before the tenth century with the founding of Ile-Ife, but reaching a peak by the middle of the

Case study I (*continued*)

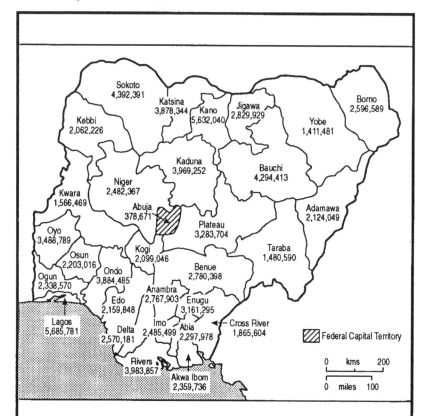

Figure 5.3 Nigeria: states and populations, 1991

nineteenth century at about the same time as the British were penetrating the country from the south. In Yorubaland are the large cities of Abeokuta, Ado-Ekiti, Ibadan, Ile-Ife, Iwo, Ogbomosho, Oshogbo, Oyo, and on the southern coast, Lagos, all with populations of over 100,000. In the south-east of the country there is another concentration of large cities such as Aba, Benin, Calabar, Enugu, Onitsha and Port Harcourt. In the north of the country Kano dominates, but other large centres include Jos, Kaduna, Katsina, Maiduguri, Sokoto and Zaria. What is notice-able from any population distribution map is the absence of many

Case study I (*continued*)

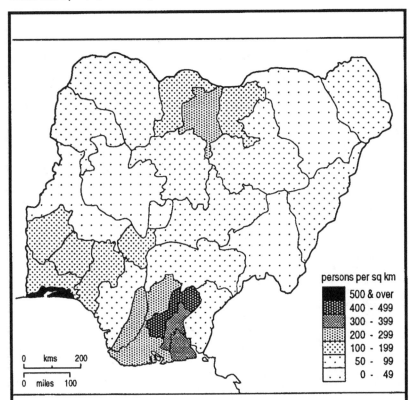

persons per sq km

500 & over
400 - 499
300 - 399
200 - 299
100 - 199
50 - 99
0 - 49

0 kms 200

0 miles 100

Figure 5.4 Nigeria: population density by state, 1991

large towns across the centre of the country and in the valleys of
the two major rivers, the Niger and Benue. This so-called 'middle
belt' has long been a relatively unproductive area of sparse population, due initially to slave raiding and the prevalence of tsetse
fly.

Since independence, national development strategies have per-
petuated a strong urban bias. Industries, infrastructure and services have been developed in the towns, whilst the rural areas have
been relatively neglected. Many Nigerians view farming as a low
priority, not the sort of work to be undertaken by anyone with an

Case study I (*continued*)

Figure 5.5 Nigeria: urban centres and communications

education. Rapid urban growth has been fuelled by a steady
stream of migrants from the countryside, although there is some
variability in trends of rural–urban migration. In-migration has
been greatest in cities such as Lagos, Kaduna, Jos and the eastern
towns of Enugu, Onitsha and Port Harcourt, but less so in some
older towns such as Katsina, Ogbomosho and even Kano and
Ibadan. There seems to have been movement out of some of the
older cities towards Lagos, some state capitals and newer cities

Case study I (*continued*)

such as Kaduna. Migrant links with the home village remain strong. In Iboland, for example, such migrants are often referred to as 'sons abroad' and are expected to maintain contact with their home and return eventually. It is common for urban-based associations, sometimes called 'Improvement Unions', to be involved in village affairs. They transmit new ideas and aspirations, are an urban lobby for village interests and provide advice and finance for village developments such as roads, bridges, schools, dispensaries and hospitals.

Rapid urban growth has caused many problems relating to housing, transport, water supply, the urban environment and urban food supply. Transport is a major problem in many Nigerian cities and Lagos's traffic congestion is legendary. Elsewhere, the situation is particularly acute in old cities like Kano, where the pre-colonial street pattern was not built for motor vehicles. Transport provision in many Nigerian cities has been left to the non-subsidized private sector and consequently buses are expensive, numbers of private motor vehicles have increased and there is little provision for pedestrians and bicycles. There is an urgent need in many cities to pedestrianize certain areas and introduce subsidized mass public transport systems to reduce road congestion.

Supplying urban populations with food is another problem. Intensive market gardening areas surround most Nigerian towns. The increasing consumption of bread as a convenience food necessitated enormous imports of wheat and flour in the late 1970s and early 1980s, amounting to 1.5 million tonnes in 1981, 90 per cent coming from USA. From the late-1980s import restrictions have been imposed, together with efforts to increase domestic wheat production, notably on irrigation schemes in the north.

The housing situation in most Nigerian cities gives much cause for concern. In Calabar, a study undertaken in 1978 showed that 50 per cent of migrants to the city had to wait longer than four months to find housing, and over 20 per cent waited longer than ten months. In the meantime, migrants shared with friends and relatives in overcrowded and deteriorating houses in the central

Case study I (*continued*)

areas. Few dwellings had electricity or flush toilets and bucket and pit latrines were common. More wealthy migrants lived in newer estates on the edge of the city. The housing situation in Lagos is particularly serious and in 1975 it was estimated that the city needed an extra 10,000 houses each year. Much of the existing housing stock is sub-standard, but its demolition, apart from being costly, would seriously disrupt the livelihood and social life of the people who live there. The growing metropolis has 'swallowed up' many former rural villages such as Olaleye, which in 1967 had 2500 people, but by 1983 had grown to 20,090 people. Three quarters of the people in Olaleye are low-income tenants renting on short leases of less than five years from landlords who are mainly self-employed businessmen, some of them absentee. Average household size ranges from 5–11 people, with 60 per cent of households occupying one room.

Since independence the private sector has continued to be the main source of housing supply in Lagos with the government playing only a marginal role. The federal (national) government did little about housing in the 1960s and early 1970s, it remaining the responsibility of state governments with little, if any, co-ordination of policies. The first successful attempt to give the housing issue a national focus was the establishment in 1971 of the National Council on Housing with members drawn from the states. A national housing programme was eventually established in 1972, aiming to build 59,000 dwellings throughout the country, with 15,000 in Lagos and 4000 in each of the states.

The Third National Development Plan, launched in 1975, declared that the federal government, '. . . now accepts it as part of its social responsibility to participate actively in the provision of housing for all income groups and will therefore intervene on a large scale in this sector during the plan period'. In 1976 the housing programme was enlarged to 202,000 units, 46,000 of these in Lagos State. Attention was to be directed towards the provision of basic infrastructure, such as water and electricity, and social amenities such as schools, health centres and shopping centres. A Mortgage Bank was established to lend to individuals, state gov-

Case study I (*continued*)

ernment housing corporations and private developers. Companies with over 500 employees were encouraged to develop housing estates for their workers. There has, however, been a wide gap between proposals and achievements due to poor planning and economic recession. By 1980 only 19 per cent of the proposed 46,000 housing units in Lagos were completed, whilst in other states the average completion was only 13 per cent, and as low as 0.5 per cent in Benue State. So, by 1980 less than one-fifth of the proposed 202,000 new housing units were completed. The generally poor record of government housing programmes raises the question that perhaps government should not undertake direct housing construction.

Rapid urban growth, and in particular the problems of Lagos, have been key factors in the decision to build a new capital city at Abuja in the sparsely settled and underdeveloped 'middle belt' region. Abuja has the advantage of being centrally located and hopefully should help to reduce the rivalry between north and south which has plagued Nigerian political history since the country was formed by the British. The master plan for Abuja was worked out from 1973 with experts from the new city of Milton Keynes in Britain (see Figure 5.6). The Federal Capital Development Authority was established in 1975 and the plan was eventually published in 1979. The new city has a semi-circular arrangement of segregated residential, commercial, industrial and government districts, linked by a network of multi-laned highways and has an international airport. Imposing government buildings will be at the centre of the arc. Residential districts of 40,000–60,000 people are grouped into 'mini-cities' of 100,000–250,000 people with associated employment facilities. Unlike many other Nigerian cities, there is likely to be far more segregation by income in Abuja. The city, which is the country's major prestige project, has a projected population of 1.6 million by 2000, with an ultimate size of 3 million. Although its construction has been more protracted than expected, the Babangida government is keen to see its completion and the movement of government ministries is likely to be completed by the end of 1993. The master plan speaks of Abuja

Case study I (*continued*)

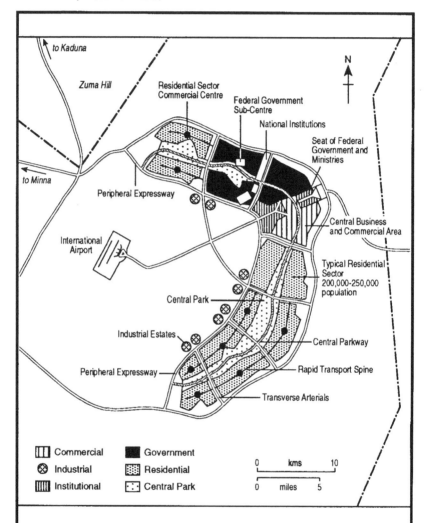

Figure 5.6 Abuja, Nigeria, the new federal capital

serving as a 'symbol of Nigeria's aspirations for unity and great-
ness, being a central, neutral and healthful place with plenty of
room for urban development'. There is no doubt that a more

Case study I (*continued*)

centrally located capital is needed and that pressure on Lagos must
be relieved. Whether the completion of Abuja will actually benefit
Lagos and other overcrowded Nigerian cities is difficult to predict
and probably rather unlikely.

Case study J

Urban employment in Ghana

In March 1957, under the leadership of the charismatic figure of
Dr Kwame Nkrumah, Ghana was the first country in Africa south
of the Sahara to become independent. The country covers an area
of 238,537 sq. km and in 1989 had an estimated population of 14.6
million, growing at the rate of 2.6 per cent per annum. Accra, the
capital city, has a population of about one million and the second
city, Kumasi, has a population of about half a million.

Ghana has had a turbulent political history since independence,
with nine governments, alternating between civilian and military
rule. The economy, which at independence was based on strong
exports of cocoa and minerals, notably gold, diamonds and man-
ganese, experienced a sharp decline in the 1970s, caused by inappropriate economic policies, escalating public expenditure and a
heavy bias towards the urban sector. Per capita GNP declined at
the annual rate of 4.4 per cent between 1973 and 1983, agricultural
and industrial output plummeted, and inflation was rampant. The
economy reached its lowest ebb in 1983, compounded by the
drought of the early 1980s, and the need to absorb 1.2 million
Ghanaian migrants who were deported from Nigeria. In that year
hunger was widespread and cocoa farmers were chopping down
their trees (once the source of the country's wealth) for firewood.

The Provisional National Defence Council (PNDC), led by
Flight Lieutenant Jerry Rawlings, seized power on 31 December
1981 from the short-lived civilian regime of Dr Hilla Limann. In
the early days, Rawlings managed to generate mass mobilization
of the people, with the establishment of local defence, price
control and anti-hoarding committees and a strong attack on

Case study J (*continued*)

profiteering and corruption – known as *kalabule*. 'People's Shops' took over the internal distribution of essential commodities. Rawlings then launched a major IMF-backed economic recovery programme (ERP) in the budget of April 1983, which reversed many of the previous government's policies with some success. The ERP was implemented in two phases, 1983–6 and 1987–9. The first phase devalued and unified exchange rates and aimed to arrest and reverse economic decline, notably in agriculture. Rehabilitation programmes were launched in the timber and manufacturing industries, in the gold fields, in transport and in the cocoa producing areas. The second phase, commencing in 1987, focused on increasing economic growth and investment and rationalized and privatized the management of many state-owned enterprises. As a result of the ERP, 1985 was the first year since 1978 in which real GDP per capita had risen and the improvement continued throughout the late-1980s. The production of food crops more than doubled between 1983 and 1986. The contribution of cocoa increased from 0.7 per cent to 8.7 per cent of GDP between 1982 and 1988 and manufacturing industry's contribution soared from 3.6 per cent of GDP to 10.1 per cent in the same period.

In June 1989 the government published its first employment statistics for ten years. In spite of a steadily rising population, recorded employment in Ghana actually fell from 482,100 in 1979 to 207,900 in 1982 and recovered to 413,700 in 1986, a figure still well below the 1979 figure. Agriculture, forestry and fishing was the most important sector, providing about 60 per cent of total employment. The second most important sector was commerce – wholesale and retail trade, restaurants etc. – accounting for about 14 per cent of total employment. Although the ERP aimed to increase the efficiency of the public sector, and many employees were made redundant in parastatal organizations, the public sector share of employment actually rose from 74.5 per cent in 1979 to 84 per cent in 1986. However, as in other African countries these statistics only relate to those who are registered with the Employment Service. A survey of employment in Ghana undertaken by the International Labour Organisation (ILO) in the late-

Case study J (*continued*)

1980s detected widespread underemployment in towns and an acute shortage of productive employment opportunities due to the pruning of the civil service and public sector. Real wages and salaries fell by an enormous 83 per cent between 1975 and 1983 and continued to fall up to 1989, forcing many workers to seek additional sources of income. Consumer prices rose by 31 per cent in 1988, 25 per cent in 1989 and 37 per cent in 1990.

The ILO report reached the gloomy conclusion that '. . . for the next several years, virtually no leavers from the primary and junior secondary schools can expect formal sector wage or salaried employment. Employment will have to be in farming and fishing, in informal service construction, road building and production activities, or in trading.' The ILO urged the Ghana government to, '. . . place greater emphasis on substantial and sustained generation of productive employment in both the modern sector and rural and informal sectors'. Other than farming, the urban informal sector is likely to have to absorb large numbers of school leavers and possibly older people made redundant from public sector organizations. Ghana has a very active and important urban informal sector, providing 65 per cent of non-agricultural employment, compared with only 15.6 per cent in modern sector employment. It is estimated that the informal sector contributes 22 per cent of total GDP. The sector is characterized by small enterprises with poor working conditions, using family labour, an apprenticeship training system and simple technology. There is considerable flexibility in movement between activities and it should be recognized that the informal sector plays an important role in the development of indigenous entrepreneurship. Emphasis in business is often on survival rather than profit, and during difficult times the informal sector plays a key role in alleviating large-scale poverty.

In Kumasi, Ghana's second city, informal sector activities include trading and manufacturing. Kumasi Central Market is the largest single market in Ghana, operating as a central node in the national food distribution network. Kumasi traders, who are mainly urban-born or urban-raised Asante women, have

Case study J (*continued*)

established strong links with a network of small-scale producers, also mainly women, who have expanded food production owing to the existence of reliable retail and wholesale outlets paying consistent prices. Entry into trading seems to be easy and there is frequent transfer between different trading roles. The traders show a remarkable degree of flexibility, responding to environmental, economic and political events with changes in their trading enterprises. When products are in short supply, traders spend long hours visiting supply villages, taking in more remote locations. Traders liaise with a regular circuit of supply areas, aiming to provide a steady flow of crop deliveries right up to the end of the harvest season when prices are higher. With the lack of space and the perishability of produce, wealthy traders with extra capital that could be used to build up stocks of food for the off-trading season often prefer to loan money to farmers to store produce on the farm rather than take the risk of storing it themselves. Some traders lend money to farmers in the months just before harvest to enable the hiring of labour for weeding or harvesting. In return, the lender has the right to buy the crop at the price prevailing when it is sold. The importance of women traders in Ghana's informal employment sector is impressive, women accounting for 84.6 per cent of total commercial employment in 1970. Professional and salaried women may also engage in part-time trading, but concentrate less on foodstuffs and more on light manufactured goods, especially cosmetics and clothing.

Over 30,000 people were employed in Kumasi's manufacturing and related repair services in the late 1970s, the bulk of these in the informal sector. Enterprises of similar type were clustered in particular parts of the city, usually in poor accommodation with 80 per cent without access to water or electricity. Unlike the food traders, many manufacturing entrepreneurs had migrated into Kumasi, mainly from other parts of Ashanti region. They were relatively well educated, with 60 per cent having more than ten years schooling and 90 per cent having received skills training through an impressive apprenticeship system. An average of 4.5 persons were employed in each enterprise, the larger businesses

Case study J (*continued*)

specializing in motor repair and carpentry and having many more apprentices. The average value of investment was $680, and the average amount of capital per worker was generally only a small fraction of that in the formal sector. The average value of output was about $160 per week. Individual earnings were higher than for equivalent work in the modern sector, and foremen or 'masters' earned higher wages when they were supervising more apprentices. The latter received token pocket money plus food. Ninety per cent of raw materials and servicing requirements were bought from retailers, usually small shops. Larger informal sector businesses had more links with the formal sector, but three-quarters of sales were to households and individuals. Though direct links with the formal sector were generally weak, indirect links through middlemen were common. For example, carpentry enterprises bought wood from merchants, who in turn had purchased from sawmills. Traders gaining contracts for school furniture usually contracted informal businesses to produce the items. Although only 3 per cent of enterprises had received a bank loan, businesses seemed to be successful in overcoming the barriers to expansion.

Employment opportunities in Ghana's urban informal sector are varied and plentiful. But in the light of job cuts following the implementation of the ERP it is likely that the informal sector will have to shoulder much of the burden in providing new employment. The ILO has proposed that the Ghanaian government should establish a special unit to develop a clear policy and incentive scheme for assisting the informal sector. Skills training, the provision of bank credit, the allocation of legally approved sites and the strengthening of enterprise associations are some of the possible measures which might be introduced to support informal sector activities and generate more employment.

Case study K

Tourism in Kenya

Tourism is big business in Kenya. Between 1984 and 1988 the tourist trade, restaurants and hotels contributed an annual average of 11 per cent to the Gross Domestic Product, and in 1987 and 1988 tourism actually earned more for Kenya than either of the two traditional exports of coffee and tea. For most of the 1980s Kenya was one of Africa's most popular tourist destinations, attracting 5.1 per cent of the continent's and 30.5 per cent of eastern Africa's tourist arrivals. During the 1980s, Germany, the UK, the USA and Switzerland dominated tourist arrivals, but there has also been a significant growth in numbers from India, Japan, Israel, Australia and New Zealand. The 676,900 tourists who arrived in the country in 1988 contributed no less than $404.7 million to the economy. This represented a considerable growth in the twenty-five year period since 1963, when tourist arrivals were 110,200 and receipts were $25.2 million.

Tourism development in Kenya has been a post-war phenomenon. In the latter part of the colonial period an increasing number of wealthy tourists arrived, mainly by ship, to visit the newly established national parks. Annual tourist numbers increased in the 1950s from around 20,000 to 40,000. But it was after independence in 1963 that tourists were attracted in larger numbers, mainly arriving by air. The rapid growth of tourism was due to:

- better air links and airports, such as the opening of Moi International Airport in Mombasa in 1979
- increased marketing by European tour operators
- promotional efforts of the Kenya Tourist Development Corporation (KTDC)
- better internal transport enabling the development of popular two-centre beach and safari holidays.

Most of Kenya's tourist facilities are located in the southern half of the country in three main locations; along the eastern coastline around Malindi and Mombasa, in Nairobi and surrounding areas and in the up-country game parks. The latter cover 7 per cent of Kenya's land surface and it is estimated that 25 per cent of total

Case study K (*continued*)

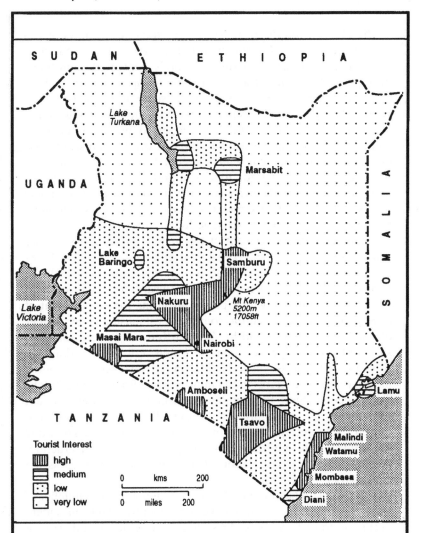

Figure 5.7 Zones of tourism in Kenya

foreign exchange comes from the direct viewing of wildlife. Over
one million visitors went to the game parks in 1988 (see Figure 5.7).

Case study K (*continued*)

The full exploitation of tourism's potential has been an obsession of successive Kenyan governments and the considerable investment in tourist infrastructure, hotels, game parks and transport has to be balanced against tourist receipts. However, as in other developing countries, only a small share of the total cost of a Kenyan holiday actually accrues to the host country. About 60 per cent of holiday costs are paid in developed countries, for air travel and to cover the administrative costs of travel agents and tour operators. Of the remaining 40 per cent a further 10 per cent flows to developed countries for tourist imports or interest transfers. Devaluation of the Kenyan currency in the early 1980s made hotels and other tourist services cheaper for foreigners, but inevitably yielded less foreign exchange for Kenya.

Approximately 78 per cent of the major coastal hotels, 67 per cent of Nairobi hotels and 66 per cent of lodges in National Parks and Game Reserves have some foreign investment. It could be argued that Kenya's 'open door' policy towards foreign investment in tourism, combined with a policy favouring the employment of local people, has enabled the country to establish a large number of good-quality hotels and has helped Kenyans to gain sufficient specialist knowledge to manage hotels efficiently. Many of Kenya's top businessmen and politicians are involved in the tourist industry and have shares in the multinational firms owning hotels. These people tend to promote tourism with government aid, sometimes when a venture is not profitable for the national economy. Private enterprise gains most revenue from tourism, while the Kenyan state subsidizes this enterprise through infrastructural facilities and the conservation of natural attractions such as the game parks.

The tourist and conference business is vulnerable in a number of ways, such as to recession in developed countries and consequent decline in tourist numbers. This happened during the oil crisis of 1972–4 and it is likely that the recession of the early 1990s will also have significant repercussions. Political instability has also affected tourism in the past, as in the mid-1970s and early-1980s when instability in neighbouring Uganda resulted in a fall in the number

Case study K (*continued*)

of tourists and hotel occupancy in Kenya in 1977 and 1982. Uncertainties about the holding of democratic elections in 1992–3 could also affect the tourist industry. There are also marked seasonal fluctuations in tourism, related to holiday periods in developed countries and climatic conditions in Kenya. Seasonality results in underutilization of productive capacity and high levels of seasonal unemployment. From April to June, for example, weather conditions are improving in Europe, but coastal Kenya experiences long heavy rains and consequently hotel occupancy rates are low.

It has been suggested that tourism generates 9 per cent of total Kenyan employment, representing 110,600 employees in 1988. However, wages of hotel staff are lower than in many other sectors of the economy except possibly agriculture and domestic service. There is tourism-related employment in industries such as soft drink and beer brewing, handicraft production, hotel entertainment, car rental, photographic studios, local restaurants and many others. There are now many linkages with local food production and importation of food for tourists declined sharply in the 1980s from 77 per cent in 1984 to only 14 per cent in 1988.

Tourists are primarily interested in seeking traditional Africa, the primitive and exotic, and many have stereotypical views of Africa and Africans. Cultural performances and traditional handicrafts emphasizing curious and primitive aspects of the traditional culture have become a vehicle for boosting the tourist industry. With the clash of cultures, the display of wealth, leisure and different moral standards, tourists can provoke and demoralize a local population. Young Kenyans living near tourist resorts often try to imitate visitors' clothes and behaviour. Also, because of greater wealth and education, the tourist is usually in a dominating position, while the local resident is in a subordinate situation. This can generate resentment among local people and may lead to open hostility. Young people sometimes become involved in semi-legal and illegal activities as 'beach boys', petty traders and beggars. Money changing, drug selling, prostitution and thieving are other such activities. The effects of tourism on local society and

Case study K (*continued*)

culture vary geographically, but are perhaps greatest in coastal Kenya where most tourists take their holidays, where the Swahili culture prevails and where the moral ethics of Islam still pervade local society.

During the 1970s and 1980s, much of the development of beach-based tourist facilities happened without any planning controls, resulting in uncontrolled ribbon development and severe pressure on facilities and local society and environment. Sea pollution from beach hotels and pressure on natural resources such as drinking water, land and marine ecology are other environmental effects of the rapidly growing tourist industry.

The effect of tourism on wildlife is another important consideration. It is in the interests of the tourist industry to protect wildlife and the Kenyan government uses this argument as its main justification for maintaining national parks and game reserves. However, farmers and herdsmen living near the game parks are increasingly unhappy about damage to crops and loss of their animals due to wild animals. Although such damage is, in principle, financially compensated by the government, compensation is often inadequate and irregular. With such a rapidly growing population, pressure on land is considerable and there is often a fundamental conflict of interests between the domestic farming and livestock economy and preservation of game parks for tourists. The traditional pastoral economy of the Masai, for example, has been severely constrained. Some critics have questioned the profitability of allocating large areas of Kenya's land for wildlife-based tourism. Future planning must aim to balance the needs of indigenous people with those of the ever-growing numbers of tourists.

Heavy reliance on foreign capital and expertise has biased Kenya's tourism industry in favour of the luxury market, much to the advantage of the foreign entrepreneurs who control the industry. Recent government policy, however, has tried to popularize domestic tourism and to increase the 'Kenyanization' of the industry with more local ownership and management. But this process of indigenization must proceed cautiously since an over-

Case study K (*continued*)

rapid replacement of expatriates could damage the country's favourable reputation as a tourism destination. Careful land-use planning will also be needed in the future, in which tourism development zones are designated after the evaluation of competing demands and possible environmental and social effects.

The continuing strength of the Kenyan tourist industry will ultimately depend on the nature and flexibility of contractual relationships between Kenyan-based firms and those in the industrialized countries from which most tourists originate. Although the tourist industry is a key element in the Kenyan economy, some would strongly criticize it on the grounds that it has done little to raise general living standards and has caused significant socio-cultural tensions in some areas.

Key ideas

1 Urbanization in tropical Africa has generally been recent and rapid, though in certain parts of the region well-established towns existed in the pre-colonial period.
2 The colonial period saw the creation of many new planned towns in which there was often a clear spatial division according to function.
3 In the past, rural and urban have been seen as quite separate, but in reality there are many complex interrelationships between them.
4 At least a third of Africa's urban population live in what might be termed 'slum' or 'squatter' settlements and these are growing at the rate of 15 per cent per annum. Urban authorities lack the funds to replace such housing with low-rental public housing units.
5 In understanding urban employment in tropical Africa, the formal/informal sector dichotomy might be criticized for being too simplistic and rigid.
6 Tropical Africa's industrial performance since the early 1960s has been very disappointing, and its share of world industrial production has actually declined in the last twenty years.

6
What future for tropical Africa?

The present predicament

Much of tropical Africa has not actually been developing in the last twenty years. On the contrary, many countries have experienced a marked deterioration in their national economic situation and in the living standards of the majority of their people. Why has this happened? What has gone wrong? These questions have been the subject of much writing and speculation and many different viewpoints have been presented. There are those who blame the legacy of colonialism, others who focus on the effects of the mid-1970s oil crisis, others who stress problems of political and economic mismanagement, others who criticize multinational companies and the imposition of structural adjustment programmes and, last but by no means least, others who believe environmental factors are the root cause of Africa's current predicament. Detailed examination of the experiences of a number of countries would reveal that in fact there are complex reasons for the current difficulties and it would be unwise to single out any factor.

That there is a crisis in tropical Africa has been frequently stated since the mid-1980s. Lloyd Timberlake commented in 1985:

> Africa's plight is unique. The rest of the world is moving 'forward' by most of the normally accepted indicators of progress. Africa is moving backwards. . . . The continent's living standards have been declining steadily since the 1970s. Its ability to feed itself has been deteriorating since the late 1960s.
>
> (Timberlake, 1985, *Africa in Crisis*, London, Earthscan)

This statement certainly seems to be supported by available statistical evidence, since in most tropical African countries the growth rate of Gross Domestic Product in recent years has been far less than the rate of population growth. Ranked alongside other countries, most tropical African states in 1985 had a per capita income of less than $1200, compared with over $7400 for the rich industrialized countries. The majority of countries with per capita incomes below $450 were in tropical Africa, where incomes declined by 4 per cent between 1970 and 1984. Further deterioration in the late 1980s is reflected in the case of Nigeria which, according to World Bank figures, moved from being the fifty-fourth poorest country in 1985, in the group of 'middle income' countries, to being the seventeenth poorest country in the 'low income' category by 1990. The position of other African countries similarly deteriorated. The region fared equally badly in a series of social variables, such as life expectancy, infant mortality and primary school enrolment.

Statistics indicate that the decline in Africa's per capita food production started as early as the 1960s. Whereas Asia experienced a marked upward trend in food production between 1971 and 1984, Africa's trend was in the opposite direction, particularly in the early-1970s and early-1980s, both of which were periods of drought (see Figure 4.1, p. 81). Additionally, between 1975 and 1985 cereal imports to Africa rose by 117 per cent and food aid by 172 per cent. These disturbing trends happened in spite of (or were perhaps exacerbated by) a large number of agricultural development projects and much advice from overseas 'experts'. In fact, advising Africa has become big business, with at any given moment, some 80,000 expatriates working for public agencies under official aid programmes. More than half of the $7–8 billion spent yearly by aid donors goes to finance these people.

There is not space here to examine all dimensions of tropical Africa's 'crisis', if indeed they could actually be identified. Attention will be focused instead on selected internal and external influences, which may have contributed to the crisis and which should receive greater priority in future development programmes.

Political and economic stability

Civil wars, military coups, widespread corruption and the lack of democratic rule have done much to deflect many tropical African countries from the development path. The state in tropical Africa is weak and fragile and political instability seems to be endemic. By mid-1984, the forty-three independent countries had suffered at least sixty successful coups, and there have been many more since 1984. The bloody Nigerian Civil War from 1967–70 led to many deaths when the former Eastern Region attempted to secede as the Republic of Biafra. Elsewhere, there have been civil wars in many countries, for example, Angola, Ethiopia, Chad, Liberia, Mozambique, Somalia and Sudan. In some cases people have suffered from both political and economic instability together with severe repression as in Uganda, Zaire and Liberia. The situation in Somalia by mid-1992 was completely anarchic, with a battered and bewildered population struggling to feed itself whilst heavily armed rival groups were fighting amongst themselves.

Uganda, a country which at independence in October 1962 was thought to have good economic potential, has had no fewer than eight different regimes in thirty years, each characterized by varying degrees of repression. The military coup of February 1971 led by Idi Amin was followed by one of the most notorious regimes that Africa has witnessed. In 1972 Amin expelled all 50,000 Asians from the country, many of whom were businessmen, technicians or managers. They left behind over 5000 businesses, many of which were re-allocated to Amin's personal friends. Declining tax revenue and enormous increases in military expenditure threw state finances into disarray, resulting in the collapse from the mid-1970s of the health and education systems, together with the entire infrastructure. Industrial production fell by 73 per cent and trade by 48 per cent whilst Amin proceeded relentlessly to eliminate any opposition to his regime. Eventually, after widespread condemnation, and following an invasion by Tanzanian troops, Amin fled the country in 1979. Since then, successive leaders have set about trying to reconstruct the fabric of Uganda's economy and society.

Civil strife has driven many millions of Africans into exile, such that in the late 1980s it was estimated that more than one-third of the world's refugee population was from Africa south of the Sahara. The major refugee movements have been from Mozambique into Malawi, and into Sudan from Eritrea and Tigray Provinces in Ethiopia and also from Chad and Uganda. The pressures on Sudan since the early 1980s have been considerable. Not only has Sudan accommodated thousands of refugees from other countries, but it suffered severe drought in the early 1980s and has its own civil war, with the Arab and Muslim north trying to impose the Islamic religion and law on the non-Islamic Southerners.

Instability and state weakness have been key factors preventing long-term development in tropical Africa. Instability interrupts development planning and reduces confidence amongst overseas investors. In an extreme form, as in Uganda, prolonged instability can lead to the complete destruction of a country's infrastructure and resource base.

One-party states and the move to democracy

There are those who believe that the absence of democratic rule in many African states has militated against progress and development and is a major cause of instability. At independence, African states were handed constitutions largely modelled on those of the colonial powers. British territories were left with Westminster-style constitutions, which the leaders of some newly independent countries felt were alien and inappropriate. Constitutional changes were often made and Kwame Nkrumah, Ghana's first leader, attempted to devise a wholly Ghanaian constitution. Julius Nyerere of Tanzania said in 1963:

> The European and American parties came into being as the result of existing social and economic divisions – the second party (that is, the Opposition) being formed to challenge the monopoly of political power by some aristocratic or capitalist group. Our own parties had a very different origin. They were not formed to challenge any ruling group of our own people; they were formed to challenge the foreigners who ruled over us. They were not, therefore, political 'parties' – ie factions – but nationalist movements. And from the outset they represented the interests and aspirations of the whole nation.

One-party states have been justified by leaders as promoting unity in striving for development and it was often argued that it was inappropriate to have an institutionalized opposition which merely criticized and retarded development measures. Governments believed they knew the best path for development and the urgency for that development necessitated a minimum of hindrance. Before long many countries had become one-party states, The Gambia being a notable exception in West Africa in preserving its multi-party system. The general experience throughout tropical Africa was first the consolidation of party power, then the establishment of a single-party state followed by, in over half the countries, the intervention of the military through a coup. In some countries the one-party state evolved peacefully, as in Senegal where after six years out of power, the opposition party merged with Leopold Senghor's dominant Union des Partis Socialistes (UPS). In

Kenya also, the opposition Kenyan African Democratic Union amalgamated with the ruling Kenyan African National Union.

Ideologies of ruling parties varied from left-wing regimes such as Guinea, Benin, Congo, Angola, Mozambique, Guinea-Bissau and Ethiopia to more right-wing capitalist approaches as followed by Cameroon, Ivory Coast, Kenya, Malawi, Nigeria and Zaire. This latter group, with the exception of Zaire, have experienced above-average economic growth, but their policies have generated more internal inequality. Summarizing the varied character of one-party rule, one might suggest it has been,

> . . . the means to govern society relatively benevolently – by Julius Nyerere, Kenneth Kaunda and Felix Houphouet-Boigny in Tanzania, Zambia and Ivory Coast respectively; more harshly but still responsibly – by Kamuzu Banda and Thomas Sankara in Malawi and Burkina Faso; to venally plunder it – as have Mobutu Seke Sese and Samuel Doe in Zaire and Liberia; or as a camouflage for personal or class tyranny – as under Jean-Badel Bokassa, Mengistu Haile Mariam or Macias Nguema in the Central African Republic, Ethiopia and Equatorial Guinea. But nearly universally the single party system has degenerated into a form of oligarchic patrimonialism that was unknown in pre-colonial Africa.
>
> (Decalo, 1992, 'The process, prospects and constraints of democratization in Africa', *African Affairs*, vol. 91, pp. 7–35)

After the collapse of the Berlin Wall and the disintegration of communism in Eastern Europe from the late-1980s, tropical Africa in the early-1990s has experienced a widespread groundswell of opinion in favour of greater democracy. Some would argue that the demise of the one-party state was already on the agenda well before the upheavals in Eastern Europe, since Africa was at a political dead-end and economically bankrupt. Economic decline was certainly a key factor in motivating the search for an alternative political system and an end to corruption. Furthermore, there was much concern about the growth and cost of the public sector, which by 1989 was consuming 60–80 per cent of national budgets. Such expenditure, combined with the inadequate provision of basic needs and social services, led to understandable impatience amongst people and a decline in respect for national political leaders. These feelings were well demonstrated in the Ivory Coast, when President Houphouet-Boigny embarked in the late-1980s on a massive and prestigious project of building a basilica in his home town of

Yamoussoukro. The Pope postponed its consecration for some time, aware of mass discontent over housing and service provision in Abidjan.

One-party states which were financially or militarily dependent on the Soviet Union collapsed first, and then France announced political conditionalities on its aid. At the meeting of Francophone countries in La Baule in June 1990, a concluding declaration stressed 'the need to associate the relevant population more closely with the construction of their political, social and economic future'. By June 1991 only a few countries, such as Benin, Ivory Coast and Gabon, had gone some way towards political reform. Houphouet-Boigny completely outmanoeuvred the pro-democracy movement by legalizing all political parties and agreeing to demands for elections. President Bongo in Gabon adopted a similar strategy. Other countries moved more slowly towards democracy, and with some reluctance amongst leaders such as Kaunda in Zambia. Pro-democracy pressures have developed throughout Africa and military rulers have fared particularly badly as in Mali (Moussa Traore), Benin (Mathieu Kerekou) and Ethiopia (Mengistu Mariam). Nigeria's military Head of State, General Ibrahim Babangida, announced a return to civilian democratic rule in 1993, after much debate over the number and nature of the new political parties. There has been concern in Nigeria and elsewhere that voting along ethnic lines should be avoided as far as possible, though in Ivory Coast and Benin this has actually happened.

By moving towards democracy the states of tropical Africa are in effect returning to where they were at independence thirty years ago. It remains to be seen whether the democratic experiment will succeed at a time of great economic difficulty. Will democracy produce true development, or will it lead to further disillusionment and perhaps sooner rather than later a return to military rule and the one-party state?

Agricultural and rural development

Political will and appropriate political structures are vital if tropical Africa is to achieve meaningful development in the years ahead. But governments must address issues which are likely to have most impact on the living standards of the bulk of the population. Agricultural and rural development, therefore, should receive top priority, since in addition to providing food, the agricultural sector generally employs the most people and its development could yield foreign exchange earnings through agricultural exports and food processing and agro-based industries.

During the 1960s and 1970s food production in tropical Africa received inadequate investment, research and development, suffered from inadequate infrastructure and commanded poor prices. While agriculture accounts for up to 60 per cent of GDP and up to 90 per cent of the labour force, the share of government expenditure going to agriculture was below 10 per cent for every country except Zambia in the first half of the 1980s. This resulted in poor agricultural performance with adverse consequences for food availability, nutrition, balance of payments and living standards. Most analyses show that poverty in tropical Africa is greatest in terms of both numbers and severity in the rural areas. Agrarian-focused development strategies must ensure full participation, maximize rural linkages and should be predominantly smallholder-based. Smallholder farms can earn an adequate living for most households and have the potential to produce marketable surpluses of food, export crops and livestock. The Kenyan government has mobilized smallholders as the focus of its agricultural development strategy. Farmers with less than two hectares of land, representing three-quarters of all farms, increased their share of national farm production from 4 per cent in 1965 to 49 per cent in 1985. In devising agricultural and rural development strategies, it is important to recognize the diversity and complexity of African agriculture and to develop country- or region-specific strategies based on natural and human resource endowments. In so doing, particular attention should be given to improving the food security of the poor.

Education

Investment in human capital, notably through education and health care, is vital if basic needs are to be fulfilled and living standards improved. Adult literacy rates in many tropical African countries are still shamefully low. UNDP statistics indicate adult (i.e. those over 15 years of age) literacy levels in 1990 as low as 18 per cent in Burkina Faso, 21 per cent in Sierra Leone, 23 per cent in Benin and 24 per cent in Somalia. In all cases female adult literacy levels were even lower, being 9 per cent, 11 per cent, 16 per cent and 14 per cent respectively in these four countries. With 45 per cent of the population of tropical Africa under 14 years of age, the educational burden on governments is considerable and education provision has suffered in recent years from cutbacks associated with government attempts to manage the crisis. Although universal primary education (UPE) may be an admirable

Plate 6.1 Makeshift school buildings at Sankandi, The Gambia.

goal, the present financial difficulties and the expenditure already allocated to education raise questions about the feasibility of such a commitment and, indeed, about the extent and distribution of education at all levels. It may seem fairer at a time of financial difficulty to prioritize primary education and basic literacy programmes whilst restricting entry into higher levels of the education system.

Ghana has recently restructured its education system and has reduced costs and made it more relevant to the country's needs. By the 1950s Ghana probably had the highest proportion of its children in school in Africa, and this expanded further after independence. However, as the national economy deteriorated, so did the schools and the quality of education. In 1976 education consumed 6.5 per cent of GDP, but this fell to 1 per cent in 1983, rising only slightly to 1.7 per cent in 1985. Enrolments also declined slightly, despite an increasing population. Since 1985, under the Economic Recovery Programme of Jerry Rawlings' government, schools have been encouraged to make a more positive contribution to economy and society. There is now a single Ministry of Education and Culture and decentralized planning in 110 districts. The school system has been reduced from 17 to 12 years, loans have been introduced in the tertiary sector and the cost of boarding has been passed on to pupils. The curriculum is also being restructured to emphasize practical and life skills rather than academic subjects. However, a strong emphasis has been placed on expanding primary school enrolments, especially for girls.

Malawi has also experienced problems in funding its education system. Primary enrolments actually dropped by 4 per cent between 1981 and 1983, corresponding closely to a period of drought and falling agricultural output. Since 1970 primary school pupil–teacher ratios have increased from 41:1 to 63:1 and per-pupil spending has declined from $14 to $6, having a significant effect on the provision of textbooks and other school materials. With World Bank funding the Malawi government is consolidating the primary curriculum and reducing the subjects studied. The university is being encouraged to raise its staff–student ratio and boarding facilities are being phased out for some new secondary schools. But the high social status of secondary and university education, and the class interests which they manifest, will limit serious moves towards a fairer distribution of resources throughout the education system.

The expansion of female education must be given high priority in future development strategies, not least because there are strong links between education and health. Studies have revealed a strong correlation between high levels of infant and child mortality and low levels of maternal education, particularly basic literacy. Educated mothers are more likely to use modern preventive health services such as immunization. Furthermore, the World Bank has found that educated farmers achieve higher productivity levels than farmers who have never been to school. Since women in tropical Africa produce up to 70 per cent of food and are the principal managers of family nutrition it is likely that investment in women's education would also result in higher food production and improved household nutrition.

Health care

Tropical Africa's main health problems relating to fertility, nutrition and communicable diseases continue to plague the population even though technological advances have controlled or eradicated these problems elsewhere in the world. Difficulties associated with childbearing are common where women have many pregnancies and there is inadequate maternity care. Infant mortality continues to be higher in

Plate 6.2 Children taking a drink from the school pump at Bwiam, The Gambia.

tropical Africa than in other major regions of the world. Diseases such as diarrhoea, pneumonia, malaria, measles and whooping cough are the main child-killers, often exacerbated by poor nutrition. Adults are subject to vector-borne, food-borne and environment-related diseases such as malaria, yellow fever, onchocerciasis, sleeping sickness and bilharzia. In the twenty years between 1965 and 1985, tropical Africa was the only region in the world where food consumption per capita actually fell, causing malnutrition which in turn leads to high mortality and severely reduces the effectiveness of labour inputs in farming and other occupations.

Since the World Health Organization's 1978 Alma-Ata conference, many tropical African countries have affirmed their commitment to extending primary health care to the entire population. But like education, health care is expensive and susceptible to expenditure cuts during difficult times. The emphasis in health care needs to shift from expensive urban-based hospitals to more widespread lower-level provision in the rural areas. As we have already seen, the importance of involving women in health-care education is vital. Women can play a key role in improving hygiene in the household. Immunization programmes and oral rehydration for children with diarrhoea have had significant effects in reducing infant mortality. Family planning programmes and advice on combating Aids should also be elements in primary health provision. Health care has links with other aspects of rural and urban development, such as the provision of clean water supplies which can have a major impact on health and living standards.

Transport

Improvements in transport facilities are also vital, particularly in rural areas where communities may be some distance from roads. Although many tropical African countries have railways, few have anything which might be called a rail network. There are many examples of countries with railways linking ports with mines and cash-crop producing areas, but minimal connections to other parts of the country. Most railways were built in the colonial period and have lacked maintenance and investment in track and rolling stock. Although rail transport could play a greater role in the development of tropical Africa, it is likely that improved road transport will have the greatest impact on the majority of people.

Better roads can bring communities within easy reach of health care and education and can play an important role in marketing crop surpluses. Costs of building and maintaining roads are very high in tropical countries, where heavy rainfall and soil erosion can sever a road link or demolish a bridge overnight. In recent years there has been a gradual shift from the construction of high-cost bituminized primary routes and highways towards the construction of low-cost non-bituminized feeder roads. Feeder roads can often be built by communities themselves using labour-intensive techniques, perhaps with the government supplying tools and some materials. Case study evidence has shown that feeder roads can result in better provision of agricultural inputs and increased crop production and marketing.

Structural adjustment: friend or foe?

In trying to cope with the crisis of the early 1980s, many tropical African governments have had to accept conditions that have locked them into new forms of dependent relationships with western governments and international financial organizations. Unable to service their massive foreign debts, governments have turned reluctantly to the World Bank or the International Monetary Fund (IMF) for loans given on condition that borrowers pursue economic policies acceptable to the international institutions. It is through such 'conditionality' that most structural adjustment programmes have been implemented.

The term 'structural adjustment programme' (SAP) has become a catchphrase in many tropical African countries since the early-1980s. Structural adjustment reforms advocated by the World Bank have aimed to make economies more efficient, more flexible and better able to use resources and thereby to engineer sustainable long-term growth. Many SAP proposals originated in the so-called 'Berg Report' to the World Bank in 1981 entitled 'Accelerated Development in Sub-Saharan Africa'. Between 1980 and 1989 some forty African governments turned to the IMF for balance of payments support, while about twenty received World Bank structural adjustment loans. Not since the days of colonialism have external forces been so powerfully focused in shaping Africa's economic structure and the nature of its participation in the world system.

Specific SAP policies have involved such measures as currency devaluation, major expenditure cuts and a reorientation towards agriculture, rehabilitation and maintenance. Taxes on consumer goods have been increased, wage rises restricted, price controls reduced and producer prices for cash crops raised. Government-owned assets have been privatized and attempts have been made to improve the efficiency of remaining governmental institutions. Increased competition and flexibility have been introduced into agricultural marketing. Some of these measures bear a strong resemblance to the market-oriented, private sector-led strategies introduced in Britain by the Thatcher government of the 1980s, and could have far-reaching and controversial political and social consequences.

Much recent debate has focused on the question of whether SAPs are actually working and how their progress should be assessed. Although it may be too early and there may not be the data available to fully evaluate the success or otherwise of SAPs, there has been little evidence of either a rise in GDP in tropical African countries or a raising of living standards. However, during the mid-1980s there was considerable currency devaluation amongst countries with SAPs. Ivory Coast's SAP seems to have reduced the gap between rural and urban per capita incomes in the period between 1980 and 1985. Nigeria's SAP, agreed by the IMF in January 1987, involved such measures as the movement to a free-market exchange rate, a ban on imports of wheat, rice and maize and the abolition of commodity marketing boards for export crops such as cocoa, rubber and groundnuts. However, it is difficult to determine whether the improvement that has occurred in the economy since SAP is due to SAP or to other factors. Although Nigerian export crop production improved, food crop performance deteriorated after 1986, in spite of rapidly rising real food prices. The 1980s may be seen as the 'lost decade' in Africa's development process and an era of 'new colonialism' under the IMF and World Bank. Some economists would argue that the 1990s must see a decisive relaxation of the external debt constraint on Africa's development and the replacement of current orthodox SAPs with more appropriate and flexible programmes. It remains to be seen whether such changes are actually feasible given the continuing economic problems in many African countries.

International cooperation and future development

Given the weak economic and political state of many tropical African countries, it could be that the best strategy for future development lies in international cooperation in the shape of political and economic groups or federations. The idea of African countries joining together to form pressure groups to face powerful developed countries and multinational companies seems a worthwhile strategy, on paper at least. In reality, however, the success and influence of such groupings have been somewhat limited.

The ideas of Pan-Africanism and a united Africa go back a long way. The cause was most effectively championed by Kwame Nkrumah who, after leading Ghana to independence in 1957, embarked on a campaign for promoting African unity. A month after French Guinea became independent in October 1958, Nkrumah formed a Ghana-Guinea 'union', and in December 1960 he invited Mali to join the union, which he saw as the nucleus for a much larger Union of African States. In January 1961 these three states met at Casablanca with the leaders of Egypt, Morocco and the Algerian Front de Liberation, as well as a representative of the Libyan government. They established a consultative assembly which became known as the Casablanca Bloc. Meanwhile, other countries were also working towards unity, with Nigeria leading a group known as the Monrovia Bloc after the Liberian capital where it first met. The former French colonies had a separate organization of their own, called the Brazzaville Bloc, with President Houphouet-Boigny of the Ivory Coast in the lead.

It was in May 1963 that these various groups came together in Addis Ababa to draw up a charter for African unity and the Organization of African Unity (OAU) was formed. Although individual African states may be quite powerless in world politics, the collectivity of the fifty-one members of the OAU, the largest regional organization in world politics, has definitely more status and influence. It is a corporate body, deriving its power and authority from its members, with no autonomy of its own and unable to act independently. A major weakness, however, is that member states have not found it easy to delegate their individual or collective powers to the OAU and the organization's role in the international system has therefore been limited. The inaugural meeting of the OAU in 1963 affirmed a policy of non-alignment with regard to all blocs. But there are many contradictions, with individual countries pursuing divergent goals in their foreign relations. Many of the francophone states, for example, have retained French military bases so as to secure larger economic benefits from France. Some African states offered military facilities to the superpowers, such as when Cuban troops were drafted to Angola and naval facilities were provided to the USA by Kenya and Somalia. The OAU Charter contains no provision for collective security in Africa and member states are not legally obliged to come to the assistance of other members in the event of aggression. There is also no standing peace-keeping force under the command of the OAU. Thus OAU responses to foreign intervention in African conflicts have reflected the attitudes of African leaders, and the OAU has been unable either to aid the victims of external intervention or to stamp out such instances. The OAU has been quite ineffective in the conflicts of Southern Africa, the Horn of Africa, Chad, Western Sahara and elsewhere.

Africa is the weakest and most marginalized region in the global system and the OAU has placed much emphasis on the need for a New International Economic Order (NIEO). The tenth anniversary summit in Addis Ababa in 1973 adopted the 'African Declaration on Co-operation, Development and Economic Independence' and put forward a demand for a NIEO, subsequently endorsed by the United Nations General Assembly in 1974. Perhaps the most significant OAU initiative in the economic field was the adoption in 1980 of the 'Lagos Plan of Action', setting out the actions which African countries should take at the national, sub-regional and regional levels to achieve economic integration by the year 2000. The OAU was also successful in persuading the United Nations to hold the first ever Special Session of the General Assembly in May 1986, exclusively devoted to the problems which Africa faced at the end of the period of severe drought. Despite these limited achievements, Africa has been unable to influence the international system in any meaningful way since the end of colonialism. As a result some would argue that a period of neo-colonialism has followed independence, supported by indigenous elites in most African states. The OAU has been unable to persuade outside powers to keep out of Africa's affairs.

In addition to the OAU, a number of other important African groupings, past and present, include:

The East African Community (EAC) EAC comprised Kenya, Tanzania and Uganda, and was formed after independence in 1961. It collapsed in 1977 due to political differences, mainly due to the Amin regime in Uganda.

The Communauté Economique de l'Afrique de l'Ouest (CEAO) CEAO was founded in 1973, and includes seven francophone West African states (Ivory Coast, Niger, Burkina Faso, Mali, Mauritania, Senegal and Benin). It has been successful in promoting trade and integration, since it has a common currency, the CFA franc, which is tied to the French franc. Regional projects are funded in the poorer countries.

The Economic Community of West African States (ECOWAS) ECOWAS was founded in 1975, and is the largest, most ambitious and richest grouping, with a population of 180 million and a GDP of $78 billion (see Case study L).

The Southern African Development Coordination Conference (SADCC) SADCC was established in 1980 by nine frontline states (Angola, Botswana, Lesotho, Malawi, Mozambique, Swaziland, Tanzania, Zambia and Zimbabwe) to reduce their extreme dependence on South Africa by strengthening regional cooperation. Priority has been given to developing transport and infrastructure and promoting industrial investment and agricultural trade.

Finally, we will examine links between tropical African countries and the European Community (EC) which have developed through the Lome Conventions. The first Lome Convention was signed in 1975 by nine (now twelve) EC countries and forty-six (now sixty-six) African,

Caribbean and Pacific (ACP) states, and was hailed as a step on the road to a new international economic order. There have been four Lome Conventions to date, the fourth having been signed in December 1989. Lomé grew out of the Yaoundé Convention when the EC enlarged and Britain became a member in 1973, necessitating the establishment of a relationship with developing countries that had lost Commonwealth preference. Under Lomé almost all ACP exports are allowed free access to the EC market, except for those products covered by the EC Common Agricultural Policy. In addition, the EC offers preferential treatment to ACP exporters of such products as sugar, beef and bananas. Stabex was established, a scheme for the stabilization of export earnings which reimburses countries suffering loss of export earnings for a variety of reasons such as a fall in prices. Stabex offers insurance cover for a number of important products such as cocoa, coffee and palm oil. Lomé also encourages investment in the industrialization of ACP states and the transfer of technology. The EC finances development projects in the ACP states, much of this aid being related to structural adjustment policies. Many ACP states have been disappointed with Lomé since their terms of trade with the EC have deteriorated. Certain important products are not covered by Stabex and in 1981–2 the compensation fund proved inadequate. Another criticism of Lomé is the meagre amount of development aid and the paternalistic way in which aid is administered. In the years ahead there is concern among ACP states that the single European market will result in 'fortress Europe' and increased protectionism. Furthermore, the changes in Eastern Europe have led ACP countries to worry that they may be further marginalized as the EC strengthens East–West relations and neglects North–South links.

With the deterioration of the economic situation in tropical Africa in the 1980s, the region's countries have been further marginalized in the world economic system. Increasingly, they have been forced to go begging to organizations such as the World Bank, the IMF and the EC. The already existing pattern of unequal exchange has been reinforced by such organizations, and the countries of tropical Africa seem unable to find a viable alternative strategy.

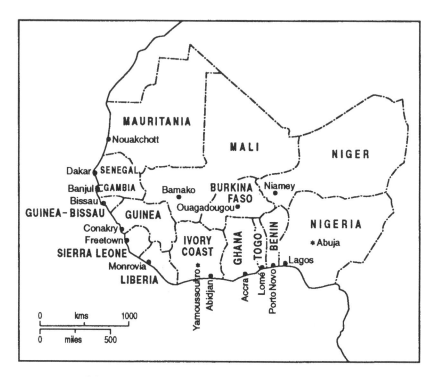

Figure 6.1 ECOWAS member countries

Case study L

The Economic Community of West African States (ECOWAS)

The Economic Community of West African States (ECOWAS) came into being on 28 May 1975 with the signing of the Treaty of Lagos by the following fifteen states; Benin, Burkina Faso, The Gambia, Ghana, Guinea, Guinea-Bissau, Ivory Coast, Liberia, Mali, Mauritania, Niger, Nigeria, Senegal, Sierra Leone and Togo. Cape Verde, the sixteenth state, joined in 1977.

Several successful and unsuccessful attempts have been made at creating economic unions in West Africa, but ECOWAS, in integrating the entire region, is the latest and most significant effort.

Case study L (*continued*)

The establishment of ECOWAS was proposed in April 1972 by Nigeria's head of state, General Yakubu Gowon and President Gnassingbe Eyadema of Togo. Initially, economic cooperation between their two countries was envisaged, but discussions with other West African leaders led to the drawing up of a treaty for regional cooperation.

West Africa covers some 6.1 million sq.km, has a population of 195 million and considerable mineral and agricultural resources. Yet it is a poor and underdeveloped region where agricultural products provide the main source of foreign exchange and the annual gross national product per capita (GNP) is less than $700 in all countries except Ivory Coast and Cape Verde. The region includes a number of small states such as Cape Verde, The Gambia and Guinea-Bissau and three landlocked states, Burkina Faso, Mali and Niger, the latter being the largest state in West Africa with 1,267,000 sq. km, much of it uninhabited desert and semi-desert.

The countries of West Africa were colonized by the British, the French and the Portuguese. They have very different traditions and there are some longstanding rivalries and boundary disputes between certain states and ethnic groups. Trade is still predominantly with the former colonial powers and less than 10 per cent of official trade is between the states of West Africa themselves. These countries are also heavily reliant on external loans from industrialized countries and the debt burdens of states such as Ivory Coast and Nigeria are quite considerable.

Nigeria dominates West Africa with a population and gross national product exceeding those of the other fifteen members of ECOWAS combined. Fears about Nigerian domination of ECOWAS existed before the 1975 treaty and persist today. General Gowon, one of the architects of ECOWAS, was himself sensitive to the fears of other states about Nigerian domination and early negotiations considered this. The final treaty made a number of compromises to allay fears and satisfy the different aspirations of the member states.

The main aim of ECOWAS is 'the rapid and balanced

Case study L (*continued*)

development of West Africa', and its key objectives to be achieved in stages are:

- the promotion of accelerated and sustained economic development
- the creation of a homogeneous society leading to the unity of the countries of West Africa
- the elimination of all customs and other duties on trade among member countries
- the establishment of a common customs tariff and common commercial policy towards non-ECOWAS countries.

Various posts and institutions were established to administer the affairs of ECOWAS, the most important being the annual meeting of the sixteen heads of state with an annual rotation of chairpersons among the members.

Other features of ECOWAS include the following;

Fund for Cooperation, Compensation and Development   The aim of the fund is to finance regional development projects and compensate member states suffering losses as a result of trade liberalization and location of community enterprises. Contributions to the fund are based on gross domestic product and per capita income of member states and vary considerably from Nigeria, which in 1977 paid 32.8 per cent of total contributions, to Cape Verde which contributed only 1 per cent.

Free Trade and Common External Tariff Agreement   This is regarded as a key feature in protecting regional industry. Intra-community trade liberalization was to be accomplished in two phases ending in 1985. However, this has been delayed due to differences of opinion between member states and other subregional organizations such as the Communauté Economique de l'Afrique de l'Ouest (CEAO), which includes many of the French-speaking states, and the Mano River Union between Guinea, Liberia and Sierra Leone. The target date for the establishment of a single monetary zone has now been postponed from 1994 to 2000.

Case study L (*continued*)

Free movement of people   Within the ECOWAS community there is, in principle, to be free movement of persons, goods, services and capital. Visas for ECOWAS citizens were abolished in 1979. However, since then certain countries such as Ghana and Nigeria have shown concern about levels of immigration. These and other states have on occasions closed their borders, and Nigeria expelled illegal aliens in 1983 and 1985, arguing that immigrants were having a bad effect on society and the economy. Although there is general commitment to the principle of free movement, in reality it has generated many problems.

Sectoral industrial development planning   The ECOWAS treaty emphasizes that programmes are designed to facilitate smooth industrial development so as to ensure a 'uniform industrial climate in order to avoid unhealthy rivalry and waste of resources'. Special treatment is to be given to the relatively less-developed members, irrespective of economic cost, in order to prevent wider economic disparities. Participation of the private sector has been encouraged by the formation of a West African Manufacturers Association at the instigation of ECOWAS. Some industrial enterprises have already been established with capital from governments, the ECOWAS fund and external investors; for example a cement factory in Benin, a palm oil plant in Ghana, a surgical cotton wool factory in Senegal and smoked fish industries in Sierra Leone. The future development of industrial planning depends on generating a community spirit and a long-term view of the benefits of staying in the community as against the short-term costs of membership.

Transport and telecommunications   Some progress has been made in transport and communications. By July 1992, 80 per cent of the Trans-Coastal Highway between Lagos in Nigeria and Nouakchott in Mauritania (4767km) had been completed as had 78 per cent of the Trans-Sahelian Highway from Dakar in Senegal to N'djamena in Chad (4633km). A number of key bridges have also been built. The Telecommunications Priority Programme has improved links

Case study L (*continued*)

within and between ECOWAS countries and established a satellite earth station in Cape Verde.

Defence pact   The ECOWAS treaty has no specific provision for cooperation on defence, but in 1978 a protocol on non-aggression was adopted to create a 'friendly atmosphere, free of any fear of attack or aggression of one state by another'. This was the first purely political decision taken by ECOWAS. In 1980 the Protocol on Mutual Assistance and Defence was signed by all states except Cape Verde, Guinea-Bissau and Mali. This provided for units from the armies of ECOWAS countries to constitute an allied community force under a force commander. There is no standing army as such, and the ECOWAS army exists only when needed to deal with emergencies. For example, a major ECOWAS concern in the early 1990s has been stabilizing the situation in Liberia after a long and bloody civil war and the overthrow of the Doe regime. In 1990 ECOWAS established a multi-national military 'monitoring group', ECOMOG, to bring peace to Liberia and other affected areas such as the Eastern and Southern Provinces of Sierra Leone. At the Abuja Summit in July 1991, ECOWAS established a five-member committee on the Liberian crisis to monitor the non-violation of the ceasefire and the subsequent free and fair electoral process.

Although some of the major elements of the 1975 treaty have yet to be implemented, ECOWAS has become a forum for promoting greater understanding and there have been some notable achievements. At the Annual Meeting in Abuja in 1991 the Declaration of Political Principles was adopted in which members committed themselves to the democratization process and the principles of respect for human rights and fundamental freedom. However, differences of opinion between member states still exist. Political developments such as the introduction of multi-party democracy, combined with economic difficulties and structural adjustment programmes in individual states, have diverted attention from ECOWAS towards domestic issues. Progress has also

Case study L (*continued*)

been constrained by financial limitations, with problems of external funding and of member states failing to pay their financial contributions.

However, ECOWAS seems 'here to stay' and has much potential given the common desire of member states to reduce the region's dependency on the former colonial powers and other developed countries. In 1991–2, a team of eminent persons under the leadership of General Gowon, one of the 'founding fathers' of ECOWAS, examined the Lagos Treaty of 1975 and made recommendations for its revision and updating to the annual meeting of Heads of State in Dakar in July 1992. In spite of limited achievements to date, it is widely recognized that ECOWAS does have an important role to play in the future development of the West African region.

Case study M

Socialist development in Zimbabwe

Since independence in 1980, the government of Zimbabwe, under the leadership of Robert Mugabe, has stressed its commitment to socialism, yet it still presides over a largely unchanged capitalist economy with prosperous commercial farming and well-developed mining and manufacturing sectors.

Zimbabwe is a landlocked southern African state with an area of 391,000 square kilometres, slightly larger than Japan. In 1989 it had a population of 9.5 million, with an annual growth rate of about 3.6 per cent. Zimbabwe's per capita GNP of $650 in 1989 placed it well above most other tropical African countries. In human terms also, Zimbabwe has one of the best records in tropical Africa, with life expectancy rising from 45.3 years in 1960 to 63 years in 1989, adult literacy increasing from 55 per cent in 1970 to 67 per cent in 1989 and, most spectacular of all, infant mortality falling from 103 deaths per thousand live births in 1965 to 46 in 1989.

Case study M (*continued*)

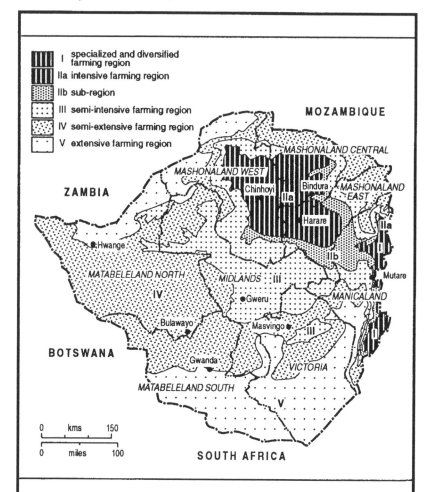

Figure 6.2 Zimbabwe: major towns and farming regions

Any study of present-day Zimbabwe must examine the colonial legacy, since the drive for independence was much more recent and traumatic than in other African countries. European coloniz-ation of Zimbabwe began in 1890 with an expedition sponsored by the British businessman Cecil Rhodes, who formed the British

Case study M (*continued*)

South Africa (BSA) Company to administer the territory. White settlers were recruited by the company, enticed with promises of land, cattle and mining concessions. The BSA Company ruled the country until 1923, when it became the British crown colony of Southern Rhodesia, ruled by a small white minority. The state played a key role in protecting white farmers and workers from African competition. Marketing boards were established for most major crops, for regulating foreign exchange and for issuing production licences. African maize producers and cattle owners were paid low prices, in effect subsidizing white farmers and their exports. Like its neighbour, South Africa, Southern Rhodesia introduced a number of apartheid-like policies, including job and residence discrimination, influx control into the white areas and pass laws for Africans. The 1930 Land Apportionment Act reserved twenty million hectares of the best land for European farmers, whilst the African majority was restricted to only twelve million hectares of overcrowded communal land. After the Second World War the economy benefited from higher mineral prices and industrial expansion, whilst white settler farming was stimulated by an increased demand for tobacco. This led to greater concentration on export crops and a further influx of white settlers, from 80,000 in 1945 to 125,000 in 1950.

In the early 1960s, when other African states were gaining their independence, the white minority government of Southern Rhodesia refused to share power with the African parties. In 1962 the new Rhodesian Front Party (RFP) came into power, under the leadership of Ian Smith, with explicit policies of protecting white privileges. Although the roots of African nationalism in Zimbabwe go back a long way, the first significant nationalist party was the African National Congress formed in 1957. This was followed by the Zimbabwe African People's Union (ZAPU) in 1962, led by Joshua Nkomo and in 1963 by the breakaway Zimbabwe African National Union (ZANU), led by Robert Mugabe. Both parties were, however, banned for twelve years from 1964 and their leaders went into exile.

In November 1965 the Rhodesian government declared

Case study M (*continued*)

independence unilaterally from Britain (Unilateral Declaration of Independence (UDI)). The international community refused to recognize the new regime and sanctions were imposed. Meanwhile, the guerilla war intensified and after their release from prison in 1975 the leaders of the ZANU and ZAPU parties formed the 'Patriotic Front', with Robert Mugabe (ZANU) leading the campaign from his base in Mozambique and Joshua Nkomo (ZAPU) from Zambia. In 1978 the Smith regime, weakened by the oil crisis, the effects of sanctions and fighting the guerilla war, signed an agreement with Abel Muzorewa and Ndabaningi Sithole which resulted in Muzorewa becoming prime minister of the state of Zimbabwe-Rhodesia. But the country remained isolated internationally, and in 1979 the British government tried to break the deadlock by convening a constitutional conference at Lancaster House in London. Agreement was reached and elections were eventually held in April 1980 with Mugabe's ZANU-PF party taking power, and Joshua Nkomo and two white RFP members joining the cabinet.

The independent government inherited a war-torn and unequal economy and society geared to serving the white settlers, who, though representing only 3 per cent of the population, earned 60 per cent of total wages. With the lifting of sanctions, GDP grew by 11 per cent in 1980 and by 13 per cent in 1981, though GDP fell sharply in 1983 and 1984 due to drought. Between 1978 and 1984 many white people emigrated, the white population stabilizing at about 120,000 by 1985. The redistribution of land was an urgent priority for the new government, which undertook to resettle landless people on commercial farmland purchased from white farmers. By 1990 some 52,000 families had been resettled on 2.7 million hectares, only about 32 per cent of the target set in the 1982–5 Transitional Development Plan which envisaged the resettlement of 162,000 families by 1985.

Zimbabwe's agricultural sector employs about 65 per cent of the labour force, producing the staple food crop, maize, and other cereal crops including wheat, millet, sorghum and barley. The main crops and animal products are sold through five marketing

Case study M (*continued*)

boards or commissions. Fears that agricultural production would collapse after independence fortunately proved unfounded. On the contrary, crop and livestock sales nearly doubled in value between 1981–2 and 1986–7. Although the country's best farmland is still mainly in the commercial sector, the share of agricultural output contributed by small-scale farmers in the communal areas rose from 9 per cent in 1983 to 25 per cent in 1988. In May 1985 producer prices for maize were raised by 28 per cent. In the same year Zimbabwe became the first African country to give food aid to Ethiopia and in 1988 the World Food Programme bought 100,000 tons of Zimbabwean grain for distribution to Malawi, Botswana and Mozambique.

In the export crop sector, tobacco contributes about 66 per cent of agricultural export revenue and employs 12 per cent of the workforce. With greater smallholder interest, tobacco hectarage and production levels have increased and the crop marketed in 1988–9 was of exceptionally high quality. Coffee and sugar production have also increased. In the livestock sector, milk production is increasing and in April 1987 producer beef prices were raised by 38 per cent. Zimbabwe is the only country in tropical Africa to export beef to EC countries. The apparent success of the commercial and smallholder farming systems operating in parallel since independence has sometimes been cited as justification for not extending the resettlement process further.

Mining was the original purpose of the country's white settlement and its expansion after independence has generated foreign exchange and stimulated demand in manufacturing industry. Gold, nickel, coal, asbestos, copper and chrome are the most important minerals in order of production value. Over 90 per cent of production is exported, and in 1990 minerals generated 23 per cent of the country's export revenue. Multinational companies dominate mining in Zimbabwe. Gold production increased during the 1980s due to the opening of several new mines, notably the Renco mine in the south-east of the country owned by Rio Tinto Zinc (RTZ). In 1989 a gold refinery was opened in Harare, which is expected to reduce Zimbabwe's dependence on South Africa. In

Case study M (*continued*)

addition to RTZ, the UK Falcon group also mines gold, whilst the asbestos industry is monopolized by UK-based Turner and Newall and chrome production is dominated by the Anglo-Amerian Corporation of South Africa and US-based Union Carbide. A Minerals Marketing Corporation was established in 1983 to market minerals internationally, but marketing patterns have changed little since independence. The medium-term outlook for Zimbabwe's mining sector is encouraging, assisted by the government's decision in October 1990 to allow companies to retain 5 per cent of the value of their export earnings to finance imported inputs.

Zimbabwe has one of the largest, most diversified and best integrated manufacturing sectors in tropical Africa, accounting for 26.4 per cent of GDP in 1990. The main industries are basic metals and metal products, food, drink and tobacco, textiles and chemicals. Between 1965 and 1982 the number of industrial products made in Zimbabwe increased ten-fold, due to a complete ban on imports of products which Zimbabwe could produce itself, and a ban on remission of profits which forced foreign companies to re-invest. Protection continues and, through the Industrial Development Corporation, the government has increased state participation in manufacturing by taking substantial shares in certain enterprises. Tourism has grown rapidly and some new high-quality joint venture hotels have been opened. But there is concern that the growth of tourism should not damage the environment.

Considerable progress has been made since independence in education and health. Before independence the white minority was much better educated than the African majority. Racial integration in schools was achieved within two years of independence. Free primary education was introduced and enrolments increased almost three-fold between 1979 and 1986. The growth of secondary enrolment was even more spectacular, with a six-fold increase due to a commitment to provide all primary school graduates with a four-year secondary course. Teacher shortages were, however, a major problem.

In 1980 infant mortality was 14 per 1000 for whites and 140 for

Case study M (*continued*)

rural blacks. Health policy after independence focused on primary health care and, as early as September 1980, free health care was introduced for all people earning less than about $75 per month. By 1987 provincial and district hospitals had been upgraded and over 200 rural health centres completed. Major campaigns have targeted child immunization and the improvement of water supply and sanitation. Programmes for family planning education and training village health workers have also been introduced.

In just over a decade since independence, Zimbabwe seems to have made great progress in social and economic terms, yet it is perhaps too early to judge the record. What is evident is that, despite the regime's commitment to Marxism-Leninism, there has been no massive socialist transformation so far. Some would argue that Mugabe's socialist approach has been rather half-hearted and many features of both economy and society, such as land distribution and industrial investment, retain capitalist and pre-independence characteristics. The years ahead may well see a more profound change in Zimbabwe's political direction.

Key ideas

1 In the last twenty years many tropical African countries have experienced a marked deterioration in their national economic situation and in the living standards of the majority of their people.
2 The post-independence period in many countries has been characterized by political and economic instability.
3 Although the agricultural and rural sectors involve the bulk of tropical Africa's people and make an important contribution to national economies, these sectors have frequently been neglected in favour of urban-biased development strategies.
4 Education and health-care systems in many tropical African countries closely resemble those introduced during the colonial period and may be inappropriate for the needs of people and nation in the late-twentieth century.
5 National development and integration are frequently inhibited by inadequate and poorly maintained transport systems which in many cases were designed to serve the colonial economy.

6 With the implementation of externally funded structural adjustment programmes, not since the days of colonialism have external forces been so powerfully focused in shaping Africa's economic structure and the nature of its participation in the world system.

7 Thus far the success and influence of regional cooperation amongst African countries have been somewhat limited, but there is considerable future potential.

Review questions and further reading

Chapter 1   Tropical Africa: continuity and change

Review questions

1 Examine some of the common images of tropical Africa and suggest how popular myths and stereotypes might be modified.
2 'Colonialism programmed African countries to consume what they do not produce and to produce what they do not consume.' Examine this statement with reference to particular colonial policies and countries.
3 Why are so many countries in tropical Africa worse off now than they were at independence?
4 Explain the difference between 'growth' and 'development' and examine the concept of 'growth without development'.

Further reading

Binns, J. A. (ed.) (1988) 'Geographical perspectives on the crisis in Africa', *Geography*, 73 (1), 47–73.

Binns, J. A. (1991) 'The Gambia: tourism versus rural development?', *Geography Review*, 5 (2), 26–9.

Griffiths, I. L. (1984) *An Atlas of African affairs*, London: Methuen.

Hopkins, A. G. (1973) *An Economic History of West Africa*, London: Longman.

O'Connor, A. M. (1991) *Poverty in Africa*, London: Belhaven.

Oliver, R. and Fage, J. D. (1988) *A Short History of Africa*, London: Penguin.

Rimmer, D. (ed.) (1991) *Africa 30 Years On*, London: Royal African Society/ James Currey.

Chapter 2   Africa's people

Review questions

1 Why are regular and reliable population censuses important but difficult to undertake in many tropical African countries?
2 What are the main factors affecting population distribution in tropical Africa?
3 Show how selected African countries illustrate different phases of the 'demographic transition'.
4 With reference to specific examples, examine some of the main causes of migration in tropical Africa.
5 Compare and contrast the indicators of human welfare in tropical African countries.

Further reading

Clarke, J. I. and Kosinski, L. A. (eds) (1982) *Redistribution of Population in Africa*, London: Heinemann.
Gleave, M. B. (1992) 'Human resources and development', in M. B. Gleave (ed.), *Tropical African Development: Geographical Perspectives*, London: Longman.
Harrison, P. (1987) *The Greening of Africa*, London: Paladin.
Hill, A. G. (1988) 'Population growth and rural transformation in tropical Africa', in D. Rimmer (ed.), *Rural Transformation in Tropical Africa*, London: Belhaven.
Timberlake, L. (1985) *Africa in Crisis*, London: Earthscan.

Chapter 3   African environments

Review questions

1 In what ways might water be seen as the key to much of Africa's survival and future development?
2 Explain the term 'desertification' and examine the evidence both for and against desertification as a widespread process in tropical Africa.
3 How useful is the concept of 'carrying capacity' in planning for future levels of human and animal populations in relation to resource availability?
4 With reference to specific food production systems examine aspects of their 'resilience' in difficult environments.

5 What might be the characteristic features of future programmes designed to promote 'sustainable development'?

Further reading

Binns, J. A. (1990) 'Is desertification a myth?', *Geography*, 75 (2), 106–13.
Lewis, L. A. and Berry, L. (1988) *African Environments and Resources*, Boston: Unwin Hyman.
Mortimore, M. J. (1989) *Adapting to Drought: Farmers, Famines and Desertification in West Africa*, Cambridge: Cambridge University Press.
Pearce, D., Barbier, E. and Markandya, A. (1990) *Sustainable Development*, London: Earthscan.
Poulton, R. and Harris, M. (eds) (1988) *Putting People First: Voluntary Organisations and Third World Organisations*, London: Macmillan.

Chapter 4   Rural Africa

Review questions

1 Why has per capita food production in Africa declined in recent years?
2 Examine some of the common perceptions of African farmers and pastoralists.
3 What are the main characteristics of traditional farming systems in tropical Africa?
4 Does pastoralism have a future in tropical Africa?
5 What roles do rural markets play in tropical Africa?
6 How might rural non-farm activities raise living standards and reduce poverty amongst rural households?
7 Should African farmers grow food crops or cash crops?
8 Describe some of the different approaches to agricultural and rural development and suggest an appropriate strategy for the future.

Further reading

Binns, J. A. (1984) 'People of the six seasons', *The Geographical Magazine*, December, 640–4.
Binns, J. A. (1987) 'Inequality and development in rural West Africa', *Geojournal*, 14 (1), 77–86.
Binns, J. A. (1992) 'Traditional agriculture, pastoralism and fishing', in M. B. Gleave (ed.), *Tropical African Development: Geographical Perspectives*, London: Longman.

Chambers, R. (1983) *Rural Development: Putting the Last First*, London: Longman.

Griffiths, I. L. and Binns, J. A. (1988) 'Hunger, help and hypocrisy: crisis and response to crisis in Africa', *Geography*, 73 (1), 48–54.

Richards, P. (1985) *Indigenous Agricultural Revolution*, London: Hutchinson.

Siddle, D. and Swindell, K. (1990) *Rural Change in Tropical Africa*, Oxford: Blackwell.

Von Freyhold, M. (1979) *Ujamaa Villages in Tanzania: Analysis of a Social Experiment*, London: Heinemann.

Chapter 5 Urban Africa

Review questions

1 Compare and contrast the pace and timing of urbanization in tropical Africa with that of other major world regions.

2 Identify the main phases of urban growth in tropical Africa and show how these are reflected in urban structure.

3 Examine the links between rural and urban in tropical Africa. How might an understanding of these links lead to a better appreciation of the development process?

4 Discuss the various strategies which might be adopted to solve the problem of urban housing.

5 How useful is the formal/informal sector dichotomy in understanding African urban employment?

6 Should industrial development be given greater priority in the countries of tropical Africa?

Further reading

Amis, P. and Lloyd, P. (eds) (1990) *Housing Africa's Urban Poor*, Manchester: Manchester University Press.

Hardoy, J. E. and Satterthwaite, D. (1989) *Squatter Citizen*, London: Earthscan.

O'Connor, A. M., (1983) *The African City*, London: Hutchinson.

Peil, M. and Sada, P. O. (1984) *African Urban Society*, Chichester: Wiley.

Potter, R. B. and Salau, A. T. (eds) (1990) *Cities and Development in the Third World*, London: Mansell.

Riddell, R. C. (1990) *Manufacturing Africa*, London: James Currey.

Chapter 6   What future for tropical Africa?

Review questions

1 To what extent might political instability be blamed for Africa's continuing problems?
2 Will multi-party democratic rule lead to more rapid development in tropical Africa?
3 What should be the main priorities in future development strategies?
4 What are the main aims and policies of externally financed structural adjustment programmes and how successful have they been in achieving their objectives?
5 Could increased international cooperation amongst African countries play a significant role in promoting future development?

Further reading

Campbell, B. K. and Loxley, J. (eds) (1989) *Structural Adjustment in Africa*, London: Macmillan.

Hodder-Williams, R. (1984) *An Introduction to the Politics of Tropical Africa*, London: George Allen and Unwin.

Onimode, B. (1992) *A Future for Africa*, London: Earthscan.

Onwuka, R. I. and Shaw T. M. (eds) (1989) *Africa in World Politics*, London: Macmillan.

Stewart, F., Lall, S. and Wangwe, S. (eds) (1992) *Alternative Development Strategies in Sub-Saharan Africa*, London: Macmillan.

Index

T - #0233 - 101024 - C0 - 216/138/11 [13] - CB - 9781138834972 - Gloss Lamination